Alex Hansen, Per C. Hemmer, and
Srutarshi Pradhan

The Fiber Bundle Model

Related Titles

Mishnaevsky Jr, L.L., Mishnaevsky, L.

Computational Mesomechanics of Composites - Numerical Analysis of the Effects of Microstructures of Composites on Their Strength

2007
Print ISBN: 978-0-470-02764-6; also available in electronic formats

Kagan, Y.

Earthquakes
Models, Statistics, Testable Forecasts

2014
Print ISBN: 978-1-118-63792-0; also available in electronic formats

Goehring, L., Nakahara, A., Dutta, T., Kitsunezaki, S., Tarafdar, S.

Desiccation Cracks and their Patterns
Formation and Modelling in Science and Nature

2015
Print ISBN: 978-3-527-41213-6; also available in electronic formats

Soumyajyoti Biswas, Purusattam Ray, and Bikas K. Chakrabarti

Statistical Physics of Fracture, Breakdown, and Earthquake
Effects of Disorder and Heterogeneity

2015
Print ISBN: 978-3-527-41219-8; also available in electronic formats

Alex Hansen, Per C. Hemmer, and Srutarshi Pradhan

The Fiber Bundle Model

Modeling Failure in Materials

WILEY-VCH
Verlag GmbH & Co. KGaA

The Authors

Alex Hansen
Norwegian University of Science
and Technology
7491 Trondheim
Norway

Per C. Hemmer
Norwegian University of Science
and Technology
7491 Trondheim
Norway

Srutarshi Pradhan
SINTEF Petroleum Research
S.P. Andersens vei 15b
7031 Trondheim
Norway

Library of Congress Card No.: applied for

British Library Cataloguing-in-Publication Data
A catalogue record for this book is available from the British Library.

Bibliographic information published by the Deutsche Nationalbibliothek
The Deutsche Nationalbibliothek lists this publication in the Deutsche Nationalbibliografie; detailed bibliographic data are available on the Internet at <http://dnb.d-nb.de>.

Print ISBN: 978-3-527-41214-3
ePDF ISBN: 978-3-527-67199-1
ePub ISBN: 978-3-527-67198-4
Mobi ISBN: 978-3-527-67197-7
oBook ISBN: 978-3-527-67196-0

Cover Design Schulz Grafik-Design, Fußgönheim, Germany
Typesetting SPi Global, Chennai, India
Printing and Binding Markono Print Media Pte Ltd., Singapore

Printed on acid-free paper

Contents

Preface *XIII*

1 **The Fiber Bundle Model** *1*
1.1 Rivets Versus Welding *1*
1.1.1 What Are Models Good For? *3*
1.2 Fracture and Failure: A Short Summary *4*
1.3 The Fiber Bundle Model in Statistics *5*
1.4 The Fiber Bundle Model in Physics *6*
1.5 The Fiber Bundle Model in Materials Science *8*
1.6 Structure of the Book *8*

2 **Average Properties** *11*
2.1 Equal Load Sharing versus Local Load Sharing *11*
2.2 Strain-Controlled versus Force-Controlled Experiments *12*
2.3 The Critical Strength *16*
2.4 Fiber Mixtures *22*
2.5 Non-Hookean Forces *24*
2.5.1 Fibers with Random Slacks *24*
2.5.2 Elastic–Plastic Model *25*

3 **Fluctuation Effects** *27*
3.1 Range of Force Fluctuations *28*
3.2 The Maximum Bundle Strength *30*
3.3 Avalanches *32*
3.3.1 The Burst Distribution *33*
3.3.1.1 The Forward Condition *35*
3.3.1.2 The Backward Condition *36*
3.3.1.3 Total Number and Average Size of Bursts *37*
3.3.2 Asymptotic Burst Distribution: 5/2 Law *40*
3.3.2.1 Asymptotic Burst Distribution: Nongeneric Cases *42*
3.3.3 Inclusive Bursts *45*
3.3.4 Forward Bursts *47*

3.3.5 Avalanches as Random Walks *49*
3.3.5.1 The Exact Random Walk *49*
3.3.5.2 Asymptotic Burst Distribution via Random Walks *51*
3.3.6 Energy Release *53*
3.3.6.1 High-Energy Asymptotics *53*
3.3.6.2 Low-Energy Behavior *55*
3.3.7 Failure Avalanches for Stepwise Load Increase *57*
3.3.7.1 Uniform Threshold Distribution *58*
3.3.7.2 General Threshold Distribution *59*

4 Local and Intermediate Load Sharing *63*
4.1 The Local-Load-Sharing Model *64*
4.1.1 Redistribution of Forces *66*
4.1.2 Determining the Failure Sequence *67*
4.1.3 Bundle Strength *68*
4.1.4 Failure of First and Second Fibers *69*
4.1.4.1 Other Threshold Distributions *75*
4.1.4.2 Localization *76*
4.1.5 Hole Size Distribution *78*
4.1.5.1 Defining the Hole Size Distribution *78*
4.1.5.2 Hole Size Distribution for Equal Load Sharing *79*
4.1.5.3 Hole Size Distribution for Local Load Sharing: Localization *80*
4.1.6 Estimating the Strength of the Local-Load-Sharing Model *82*
4.1.6.1 Lower Bound for the Largest Hole *83*
4.1.6.2 Competing Failure Mechanisms *85*
4.1.6.3 Uniform Threshold Distribution *86*
4.1.7 Force and Elongation Characteristics *88*
4.1.8 Burst Distribution *90*
4.2 Local Load Sharing in Two and More Dimensions *93*
4.2.1 Localization *94*
4.2.2 Similarity with the Equal-Load-Sharing Fiber Bundle Model *95*
4.2.3 Burst Distribution *99*
4.2.4 Upper Critical Dimension *99*
4.3 The Soft Membrane Model *101*
4.4 Intermediate-Load-Sharing Models *104*
4.4.1 The γ-Model *105*
4.4.2 The Mixed-Mode Model *106*
4.5 Elastic Medium Anchoring *107*
4.5.1 Size Equals Stiffness *109*
4.5.2 Localization in the Soft Clamp Model *109*
4.5.3 Asymptotic Strength *110*
4.5.4 Fracture Front Propagation *112*

5 Recursive Breaking Dynamics *115*
5.1 Recursion and Fixed Points *116*
5.2 Recursive Dynamics Near the Critical Point *119*
5.2.1 Universality *120*
5.2.1.1 General Threshold Distribution *124*
5.2.2 Postcritical Relaxation *125*
5.2.2.1 Uniform Threshold Distribution *126*
5.2.2.2 General Threshold Distribution *129*
5.2.3 Precritical Relaxation *131*
5.2.3.1 Uniform Threshold Distribution *131*
5.2.3.2 General Threshold Distribution *133*
5.2.4 Critical Amplitudes *134*

6 Predicting Failure *137*
6.1 Crossover Phenomena *138*
6.1.1 The Avalanche Size Distribution *138*
6.1.1.1 Burst Avalanches at Criticality *141*
6.1.2 Energy Bursts *142*
6.1.3 The Crossover Phenomenon in Other Systems *144*
6.1.3.1 Earthquakes *144*
6.1.3.2 The Fuse Model *148*
6.2 Variation of Average Burst Size *152*
6.3 Failure Dynamics Under Force-Controlled Loading *152*
6.4 Over-Loaded Situations *155*
6.4.1 Breaking Rate of Loaded Fiber Bundles *155*
6.4.1.1 Uniform Distribution *155*
6.4.1.2 Weibull Distribution *157*
6.4.1.3 Large Overload Situation *158*
6.4.2 Energy Emission Bursts *159*
6.4.2.1 Uniform Distribution *161*
6.4.2.2 Weibull Distribution *162*
6.4.2.3 Large Overload Situation *164*
6.4.3 Energy Burst Pattern *166*

7 Fiber Bundle Model in Material Science *169*
7.1 Repeated Damage and Work Hardening *170*
7.1.1 The Load Curve *171*
7.1.2 Repeated Damages *171*
7.1.3 Damages Ending in Complete Bundle Failure *174*
7.2 Creep Failure *175*
7.2.1 A Model for Creep *176*
7.2.2 A Second Model for Creep *180*
7.2.2.1 Damage Accumulation *180*
7.2.2.2 Time Evolution *181*
7.2.2.3 Healing *182*

7.3 Viscoelastic Creep *183*
7.4 Fatigue Failure *186*
7.4.1 A Fatigue Experiment *186*
7.5 Thermally Induced Failure *188*
7.5.1 Failure Time for a Homogeneous Fiber Bundle *188*
7.5.2 Failure Time for Low Thermal Noise *189*
7.6 Noise-Induced Failure *190*
7.7 Crushing: The Pillar Model *194*

8 Snow Avalanches and Landslides *197*
8.1 Snow Avalanches *197*
8.2 Shallow Landslides *199*

Appendix A Mathematical Toolbox *203*
A.1 Lagrange's Inversion Theorem *203*
A.2 Some Theorems in Combinatorics *204*
A.2.1 Basic Selection Theorems *204*
A.2.2 A Distribution with Restrictions *205*
A.3 Biased Random Walks *206*
A.3.1 Probability of No Return *206*
A.3.2 Gambler's Ruin *207*
A.4 An Asymmetrical Unbiased Random Walk *209*
A.5 Brownian Motion as a Scaled Random Walk *211*

Appendix B Statistical Toolbox *213*
B.1 Stochastic Variables, Statistical Distributions *213*
B.1.1 Change of Variable *213*
B.1.1.1 A Useful Interpretation *213*
B.1.2 The Characteristic Function *214*
B.1.3 The Central Limit Theorem *215*
B.2 Order Statistics *216*
B.2.1 Ordering the Variables *216*
B.2.2 The Average of the mth Ordered Variable *217*
B.3 The Joint Probability Distribution *217*
B.3.1 Extreme Statistics *218*
B.3.2 The Largest Element: The Three Asymptotic Distributions *218*
B.3.2.1 The Gumbel Distribution *219*
B.3.2.2 The Second and Third Asymptotes *220*
B.3.3 The Smallest Element: The Weibull Distribution *221*

Appendix C Computational Toolbox *223*
C.1 Generating Random Numbers Following a Specified Probability Distribution *223*
C.1.1 When the Cumulative Probability May Be Inverted *223*

C.1.2 Gaussian Numbers *224*
C.1.3 None of the Above *225*
C.2 Fourier Acceleration *225*

References ***229***

Index ***233***

Book Series: Statistical Physics of Fracture and Breakdown
Editors: Bikas K. Chakrabarti and Purusattam Ray

Why does a bridge collapse, an aircraft or a ship break apart? When does a dielectric insulation fail or a circuit fuse, even in microelectronic systems? How does an earthquake occur? Are there precursors to these failures? These remain important questions, even more so as our civilization depends increasingly on structures and services where such failure can be catastrophic. How can we predict and prevent such failures? Can we analyze the precursory signals sufficiently in advance to take appropriate measures, such as the timely evacuation of structures or localities, or the shutdown of facilities such as nuclear power plants?

Whilst these questions have long been the subject of research, the study of fracture and breakdown processes has now gone beyond simply designing safe and reliable machines, vehicles and structures. From the fracture of a wood block or the tearing of a sheet of paper in the laboratory, the breakdown of an electrical network on an engineering scale, to an earthquake on a geological scale, one finds common threads and universal features in failure processes. The ideas and observations of material scientists, engineers, technologists, geologists, chemists and physicists have all played a pivotal role in the development of modern fracture science.

Over the last three decades, considerable progress has been made in modeling and analyzing failure and fracture processes. The physics of nonlinear dynamic, many-bodied and non-equilibrium statistical mechanical systems, the exact solutions of fibre bundle models, solutions of earthquake models, numerical studies of random resistor and random spring networks, and laboratory-scale innovative experimental verifications have all opened up broad vistas of the processes underlying fracture. These have provided a unifying picture of failure over a wide range of length, energy and time scales.

This series of books introduces readers – in particular, graduate students and researchers in mechanical and electrical engineering, earth sciences, material science, and statistical physics – to these exciting recent developments in our understanding of the dynamics of fracture, breakdown and earthquakes.

Preface

One of the authors, Alex Hansen, first came across the fiber bundle model in 1989 in a paper entitled *Elasticity and failure of a set of elements loaded in parallel*[1] by Didier Sornette. Interesting he thought, but it is too simple to contain any of the effects seen in fracture models that were reminiscent of critical phenomena. These were still very much in the focus of research in statistical physics at that time. A year later he was invited to give a talk at the Norwegian Institute of Technology in Trondheim on the subject of statistical models of fracture. The night before the talk, he suddenly had the inspiration to program the fiber bundle model and measure the distribution of the number of fibers that fail simultaneously during breakdown, that is, the burst distribution, expecting that it would be a boring exponential in contrast to the power laws seen in more complex models. With this result he would then demonstrate that long-range correlations are necessary to produce power law distributions that signal criticality. Plotting the result well past midnight had the effect of an alarm clock. The burst distribution followed a power law! The rest of the night was spent checking the program, but nothing was wrong. The result was presented at the talk the morning after as a very puzzling result. This triggered the interest of Per Christian Hemmer. He saw that the model presented a very nice problem in combinatorics. A collaboration ensued and we realized that the fiber bundle model is in fact a great and subtle model – as Henry Daniels already had realized more than 45 years earlier. Srutarshi Pradhan came to the fiber bundle model through his PhD work with Bikas K. Chakrabarti of the Saha Institute for Nuclear Physics in Kolkata: work that demonstrated how the fiber bundle model in fact does represent a great example of a critical system.

Yes, the fiber bundle model is a great and subtle model. It is acquiring the position in the science of fracture that spin models have in condensed matter physics. The model hovers right on the border between being analytically tractable and numerically amenable. It appeals to practitioners of three different sciences:

1) Statistical physics
2) Statistics and
3) Material science.

1) J. Phys. A **22**, L243 (1989).

We have tried to cater to these three groups in the way we have ordered the material that we present. The subject is by its very nature mathematical and the reader will meet many equations. But they never go beyond the mathematics that any student majoring in any of the three disciplines will have met. Mathematics and statistics that do go beyond the usual courses we have collected in appendices. We have also added an appendix on some computational facts that are useful when implementing the fiber bundle model on the computer – and we do encourage the serious student to implement the model computationally. This builds intuition.

We start the book by discussing the concept of a model. The fiber bundle model may seem ridiculously simple to the material scientist; how can it have anything to do have anything to do with the complexities of real materials? We answer this question in Chapter 1.

Chapters 2 and 3 provide an introduction to the "classical" equal-load-sharing fiber bundle model originally introduced in 1926 by Peirce. Armed only with these two chapters, the reader will be able to follow and contribute to the current research on this model.

Chapter 4 goes beyond the equal-load-sharing fiber bundle model. This chapter is placed right at the current research front and, therefore, the story that is presented is not complete. But we go as far as is possible without getting into too deep water. This chapter might be more challenging than the others from a mathematical point of view.

Chapter 5 casts the equal-load-sharing fiber bundle model in the language of dynamical systems theory – the discipline to which chaos theory belongs. The concepts of criticality are introduced and the contact to the field of physics that deal with such phenomena is made in this chapter.

Chapter 6 addresses what signs to look for that will tell whether a fiber bundle model is about to collapse completely or not. This is a central question in a wide range of fields ranging from mechanical engineering to seismology.

Chapter 7 is geared at material scientists. We consider fatigue, creep, and other relevant concepts here.

The final chapter considers the use of the fiber bundle model in geosciences. This is a new development that is still very much in its infancy.

When writing a scientific paper, the author knows that there will come a moment of truth: when the e-mail arrives from the editor with the comments by the reviewers. Did the reviewers get what we tried to say? Did they agree? Were our arguments sound? Did they like the results? The answer to some or all of these questions may be "no" and the struggle begins; a struggle that in the end will either be an improved manuscript being published or an unsaveable manuscript being thrown out. This is as it should be. This is the essence of science: all new work is to be challenged. When publishing a book, this process is reversed. One approaches the publisher with an idea to a book. This is then sent to reviewers who make their comments. If they are positive, a contract is signed the authors get writing. Hence, the content of the book is the sole responsibility of the authors.

Writing a book is not comparable to composing a long review article. A review article needs to cover the field completely, and the authors can expect to hear from people whose work has been accidentally left out. It is not so with a book. A book is a presentation of a personal view on a field, and does not have to cover it. We have tried to stitch together a coherent text with the three communities that use the fiber bundle in mind (physics, statistics, and material science), but there are several topics that we have not touched. An example is the computational methods for fiber-reinforced materials by William A. Curtin and others. However, we have, for the material that we do cover, focused on having a text that describes the topics at sufficient depth to be self-sufficient.

There are many to thank for innumerable scientific discussions and pleasant collaborations over the years: Bikas K. Chakrabarti, E-jian Ding, Erling Fjær, Knut Skogstrand Gjerden, Jonas T. Kjellstadli, Morten Kloster, Ferenc Kun, Peter Lehmann, Subhrangshu S. Manna, Dani Or, S. Leigh Phoenix, Purusattam Ray, Stéphane Roux, Arne Stormo, Santanu Sinha among many others. They have all contributed to this book, whether they know it or not.

We acknowledge the Norwegian Research Council for having supported our research on the fiber bundle model over the years. Two grants have provided partial funding for the writing process itself: NFR CLIMIT grant 199970 and NFR INDNOR grant 217413/E20. Srutarshi Pradhan also thanks SINTEF Petroleum Research, Trondheim, for partially funding during the writing.

We are most grateful to Susan Hough, Charles Richter's biographer, for providing portraits of Richter and Beno Gutenberg, taken from her own collection of high-quality scans of Caltech Seismological Laboratory historical photographs. We also acknowledge with gratitude the help of Julia Blackwell at the Statistical Laboratory of Cambridge University. She searched for and sent us several Daniels photos. The Waloddi Weibull portrait we received from Professor Sam Saunders. We are indebted to him for, in the midst of a difficult moving process, to be able to unpack the original portrait and for having had made a high-quality digital copy for us.

Trondheim,
May 21, 2015

Alex Hansen
Per Christian Hemmer
Srutarshi Pradhan

1 The Fiber Bundle Model

Dear reader, if differential geometry is your field, please put this book back on the shelf. It is not for you. The fiber bundles that we deal with here are not spaces, but bundles of breakable fibers. Fibers that stretch and fail. They belong to the realm of engineers, physicists, and statisticians. They are models for how materials fail under duress.

Most materials do not consist of fibers. But materials are prone to failure under loading. Keeping material failure under control is one of the most important tasks of engineering. We need to be able to trust that our buildings will not collapse, our airplanes do not disintegrate in mid-air, our tankers do not rupture at sea ...

Given the variety of materials and configurations they are used in, it seems a daunting task to attempt constructing a general theory of fracture and failure. Such a theory exists, however, and goes under the name of linear elastic fracture mechanics (*LEFM*) [1]. This has become a very refined theory over the years, and there is no doubt that it has been successful. Linear elastic fracture mechanics has as a starting point the theory of elasticity. This is a theory that treats materials as continuous, and as a result, linear elastic fracture mechanics is a top-down approach.

A completely different approach has come to life over the past couple of decades: atomistic modeling [2]. This approach hinges on the advent of the computer as a serious research tool. It is now possible to model materials with (fairly) realistic forces between the atoms in such quantities that it is possible to hook the results up with those approaches that start from a continuum description: top-down meets bottom-up.

Is there then any room for simplified fiber bundle models in the middle? Our answer is yes, and we will use the next couple of hundred pages or so to convince you, dear reader.

1.1 Rivets Versus Welding

Here is a couple of examples of failures that seem to be opposite of each other in order to highlight the complexity of the central problem: how to ensure that structures do not fail.

The Fiber Bundle Model: Modeling Failure in Materials, First Edition.
Alex Hansen, Per C. Hemmer and Srutarshi Pradhan.

We zoom in on the failure of the hull of a Boeing 737 airplane during Aloha Airlines flight 243 on April 28, 1988, where a part of the fuselage of the airplane was ripped away mid-air; see Figure 1.1. Amazingly, the pilots were able to land the aircraft with 89 passengers and 6 crew members. The failure process had started long before as a small crack near a rivet due to metal fatigue initiated by crevice corrosion. The crack grew due to the cyclic pressure loading from flying and being on the ground. As the length of the crack grew, the stresses in front of it increased, and at some point, it became unstable, opening up the fuselage by moving from rivet to rivet in the way perforated paper fails. Clearly, understanding what happened and how it can be prevented from happening again belongs to the realm of engineering. However, the growth of the initial crack and how it went unstable are just as much a problem for fundamental science: what are the underlying mechanisms and how do they manifest themselves? In the AA flight 243 incident, the rivets played a crucial role.

The American *Liberty* cargo ships produced during World War II were the first ships that had hulls that were welded rather than riveted. Yet, 12 out of the 2710 ships built broke in half without warning. Cracks formed, grew slowly, went unnoticed, and at some point, they became unstable, breaking the ship apart, see Figure 1.2. By removing rivets, no mechanism was present that could lead to crack arrest. A growing crack in a car wind shield is effectively stopped by drilling a hole in front of it. The high stress at the crack tip, which drives the crack forward, is lowered as it is spread over the surface of the drilled hole when the crack reaches it. In the same way, rivets would stop growing cracks in the hull.

This contradicts the important role played by rivets in the AA flight 243 incident where rivets were the cause of the failure. Here, the lack of rivets was the reason for the failure.

Are there fundamental and general principles at work here that can explain the difference between the two incidents? The answer is yes. But, to be able to understand these principles, we need to simplify the problem. We need *models.*

Figure 1.1 The Boeing 737 after the explosive decompression that occurred during flight on April 28, 1988, in Hawaii. (Photo credit: National Transportation Safety Board)

Figure 1.2 The Schenectady after it broke into two on January 16, 1943, in dock in Portland, Oregon. The ship had just been finished and was being outfitted. The failure was sudden and unexpected.

It is here that the fiber bundle models enter. They are models that simplify the problem of failure to the point where the very powerful methods of theoretical physics, statistics, and mathematics may be fully explored. They help us understand the subtle interplay between forces and strength that control the failure process. They help us understand what is generally present in all failure processes and what is specific for a given failure process.

1.1.1 What Are Models Good For?

Since the use of models is sometimes viewed with some skepticism by the engineering community, we elaborate some more on what precisely is a model.

Fundamental sciences, and physics in particular, approach Nature in a *hierarchical way* [3]: more general questions are posed and answered before more specific questions. We may illustrate this by the following example: in the 1920s, general quantum mechanics was developed. In the 1930s, a general theory of metals was developed. This allowed for studying specific metals, but it also opened up for the search for a class of materials that were between metals and insulators: semiconductors. In the 1940s, this resulted in the construction of the first transistor – and the electronics age was born. One may only speculate how long it would have taken to construct the transistor if this path from the more general to the more specific had not been followed. How long would it take before someone accidentally stumbled across semiconductors?

This hierarchical approach lies behind the extensive use of models in theoretical physics. The fiber bundle model is a good example of the use of physical models to study the physical phenomena of interest with the minimum of ingredients needed: these models are stripped of any irrelevant contents. In fact, the models, and the approach of physics to science, are related to Occam's dictum: *Numquam ponenda est pluralitas sine necessitate* [plurality must never be posited without necessity] [4].[1)]

Still, the fiber bundle models have proved to be very effective in *practical* applications such as fiber-reinforced composites. In this context, the models have a history that goes back to the 1920s [6], and they constitute today an elaborate toolbox for studying such materials, rendering computer studies orders of magnitudes more efficient than brute force methods. Since the late 1980s [7], these models have received increasing attention in the physics community due to their deceivingly simple appearance coupled with an extraordinary richness of behaviors. As these models are just at the edge of what is possible analytically and typically not being very challenging from a numerical point of view so that extremely good statistics on large systems are available, they are perfect as model systems for studying failure processes as a part of theoretical physics.

1.2 Fracture and Failure: A Short Summary

Fracture and material stability have for practical reasons interested humanity ever since we started using tools: our pottery should be able to withstand handling, our huts should be able to withstand normal weather. As science took on the form we know today during the Renaissance, Leonardo da Vinci studied 500 years ago experimentally the strength of wires–fiber bundles–as a function of their length [8]. Systematic strength studies, but on beams, were also pursued by Galileo Galilei 100 years later, as was done by Edme Mariotte (of gas law fame), who pressurized vessels until they burst in connection with the construction of a fountain at Versailles. For some reason, mainstream physics moved away from fracture and breakdown problems in the nineteenth century, and it is only during the last 20 years that fracture problems have been studied within physics proper. The reason for this is most probably the advent of computers as a research tool, rendering problems that were beyond the reach of systematic theoretical study now accessible.

If we were to single out the most important modern contribution from the physics community with respect to fracture phenomena, it must be the focus on *fluctuations* rather than averages. What good is the knowledge of the average behavior of a system when faced with a single sample and this being liable to breakdown given the right fluctuation? This book, being written by physicists, reflects this point of view, and hence, fluctuations play an important role throughout it.

1) Einstein is often quoted as having stated "Everything should be as simple as possible, but not simpler." This is of course the Occam razor [5].

1.3
The Fiber Bundle Model in Statistics

Even though we may trace the study of fiber bundles to Leonardo da Vinci, their modern story starts with the already mentioned work by Peirce [6]. In 1945, Daniels published a seminal review-cum-research article on fiber bundles, which still today must be regarded as essential reading in the field [9]. In this paper, the fiber bundle model is treated as a problem of statistics and the analysis is performed within this framework, rather than treating it within materials science. The fiber bundle is viewed as a collection of elastic objects connected in parallel and clamped to a medium that transmits forces between the fibers. The elongation of a fiber is linearly related to the force it carries up to a maximum value. When this value is reached, the fiber fails by no longer being able to carry any force. The threshold value is assigned from some initially chosen probability distribution and does not change thereafter. When the fiber fails, the force it carried is redistributed. If the clamps deform under loading, fibers closer to the just-failed fiber will absorb more of the force compared to those further away. If the clamps, on the other hand, are rigid, the force is equally distributed to all the surviving fibers. Daniels discussed this latter case. A typical question posed and answered in this paper would be the average strength of a bundle of N fibers, but also the variance of the average strength of the entire bundle. This book takes the same point of view, discussing the fiber bundle model as a *statistical model.*

Fredrick Thomas Peirce

© Special Collections Research Center,
NCSU Libraries, Raleigh NC, USA

Fredrick Thomas Peirce was born in 1896 in Southport, Australia, where his father was a minister. He was awarded the B.Sc. degree by the University of Sidney at the age of just 19. During World War I, he served in the Australian Army Signal Service and was severely wounded in Palestine.

After the war, Peirce went to England and studied X-ray crystallography under the Nobel laureate W. H. Bragg, at University College, London, as well as other topics in physics and chemistry.

In 1921, he joined the Physics Department of the British Cotton Industry Research Association in Manchester. His background enabled him to apply the principles of physics to the study of textile structures. When a new Testing Department was created in 1924, Peirce was chosen as its head.

In the beginning, his research was centered on fundamental physical properties of cotton fibers and yarns: rigidity, plasticity, and tensile behavior. His article *Theorems of the strength of long and of composite specimens* was probably the first scientific treatment of fiber bundles [6]. Here he developed a *weak link* theory, which dealt with the variation in bundle strength due to the probability of random weak spots.

After 1924, his research focused on the influence of humidity on textiles: how to prevent water from entering from the outside while allowing it to escape from the inside. It has been suggested that this was triggered by his change of environment from the dry Australia to the damp Manchester.

During World War II, he and his staff of 60 were devoted to meeting military requirements, as textiles for parachutes, and for arctic and tropic clothing. In 1944, his expertise was requested in the United States, and he accepted an invitation to become Director of Textile Research in the School of Textiles at North Carolina State College. However, after a stroke, he resigned and moved back to Australia, where he died at an early age of 53.

1.4 The Fiber Bundle Model in Physics

Theoretical physics has changed quite profoundly over the last three decades. This coincides with the computer coming of age. We are not there yet, but computational physics is rapidly establishing itself as a third way of doing physics on equal footing with experimental and theoretical physics. The power of the modern computer, being in the form of a huge machine such as the Japanese K-computer consisting of 800 racks or in the form of GPUs – Graphic Processing Units – which allows enormous power on the desktop thanks to the computer game industry, allows for handling problems that would be forbidding even to think of in earlier times.

When looking back at the history of fracture and failure, one realizes that the early giants, for example, Galilei, struggled with such problems. However, in the eighteenth century physics moved away from such problems. They were, we suspect, too dirty (i.e., practical) to touch for pure physics. Perhaps substituting the term "dirty" by "difficult" is closer to the truth, however. The divorce between physics and the science of fracture and failure lasted until the 1980s when the statistical physics community – still quite elated after the tremendous successes in connection with critical phenomena – started considering such problems. In these early days of the renewed interest in fracture, it was the fuse model that got all the attention. This numerical model, which we will consider in Chapter 6, simulates a network of electrical fuses, each having a current threshold drawn from some probability distribution. As the current is increased through the network, how do fuses burn out? In the beginning of the failure process, fuses burn out because they have small thresholds. However, as the failure process proceeds, the current distribution in the network evolves and fuses carrying high currents appear. Hence, there are now fuses that fail not because they are weak, but because the currents they are carrying are high. A competition between these two effects starts and a rich variety of different effects may be studied. The problem with the fuse model is that it is very difficult to get any hard – mathematically derived – results. This model is a numerical model only.

The fiber bundle model in its more sophisticated version – the local-load-sharing fiber bundle model (see Chapter 4) – shows the same competition between stress enhancement and weak fibers as the fuse model does [10]. The distribution of fiber strengths makes the cracks – missing fibers – repulsive with respect to each other. This is easy to understand. If we sit on a failed fiber and look for weak fibers around it, the further away we look, the weaker the weakest fiber we find will be. On the other hand, the stress enhancement due to failure will be the highest near the failed fiber. Hence, if fibers fail for this reason, they will be near the already-failed fiber. Hence, stress enhancement makes cracks – failed fibers – attractive.

Referring back to Section 1.1, we discussed two accidents: the explosive compression of an airplane due to fractures evolving from the rivets holding the hull together and the breaking up of a tanker as a result of there being no rivets in the hull. These two accidents seem in some sense opposite of each other. However, in light of the competition between weak thresholds and stress enhancement, we may understand the principle behind the two accidents. In the case of the rivets in the airplane hull, they were weak spots where failure would occur. Cracks grew from these weak spots and at some point (April 28, 1988) the stress enhancement took control with the result that the crack growth took off. The same principle happened in fact in connection with the tanker. Even though there were no rivets, the hull would not be completely uniform and microcracks would appear. However, the competition would soon be won by the stress enhancement at the crack tips.

To our knowledge, the first use of the *fiber bundle model* in physics came in 1989 with Sornette as author [7]. In contrast to the fuse model, the fiber bundle model offers a fine balance between being analytically tractable and numerically amenable. In its simplest form, the equal-load-sharing fiber bundle model, essentially everything may be calculated. In more advanced models, for example, the local-load-sharing fiber bundle model (see Chapter 4), some quantities may be calculated analytically in these models, but others may not.

The authors of this book are physicists. This book reflects this fact. However, one of the principal tools that statistical physicists use is statistics. We will use statistics extensively in this book. It is only in the choice of which subjects to emphasize that our background will shine through.

1.5
The Fiber Bundle Model in Materials Science

The fiber bundle model has so far been presented here as a general model rather than as a tool that can be used in engineering. The fiber bundle model does have a place in engineering. Perhaps not surprisingly, it has carved out a fairly sizable field in fiber reinforced composites [11]. We will not go into detail in this field, but just point out that this endeavor began in 1952 with the Cox shear-lag model [12]. It then passed through different stages of development, for example, the Hedgepeth model [13], until now being a mature model that is used for strength calculations in an engineering setting; see, for example, [14].

In the fiber bundle model, the material failure occurs when the fibers are stretched and some may possibly fail. A similar model, in which, however, failures occur under compression, is the pillar model [15, 16]. In this, two solid horizontal planes are supported by a set of pillars. The pillars have statistically distributed thresholds for failure under compression. In their simplest versions, there is a one-to-one correspondence between the pillar and the fiber bundle models. In both models, one is free to specify how stress is redistributed around a failure. The pillar model has been used for describing several aspects of rock failure under compression.

In Chapter 7, we will say a few more words about fiber-reinforced composites and revisit the pillar model.

1.6
Structure of the Book

This monograph summarizes the authors' current knowledge of the fiber bundle model. It is written from the perspective of theoretical physics. Chapter 2 introduces the equal-load-sharing fiber bundle model and takes the reader through the "classical" results concerning its average properties. Chapter 3 focuses on fluctuations and here important concepts such as avalanches are introduced. Chapter 4 introduces the local-load-sharing model as the opposite limit of the equal-load-sharing model. Whereas the latter distributes the forces carried by the failed fibers equally among all remaining fibers, the local – load-sharing model distributes the forces to the nearest surviving fibers. Also, more sophisticated models such as the soft clamp model are discussed here. Chapter 5 returns to the average properties of the equal-load-sharing fiber bundle, but now as an iterative system. The theory of iterative maps is used to demonstrate the relation between the equal-load-sharing fiber bundle model and critical phenomena. Chapter 6 discusses the possibility to predict the point at which the fiber bundle model collapses under load, based on the theory developed in Chapter 3 on avalanches and on the iterative theory in Chapter 5. In Chapter 7, we discuss the use of fiber bundle models in connection with important phenomena such as creep, fatigue and crushing in addition to considering the influence of thermal noise on the fiber bundle. Finally, in Chapter 8, we discuss briefly the use of fiber bundle models in geophysics, particularly in connection with snow and landslide avalanches.

2
Average Properties

2.1
Equal Load Sharing versus Local Load Sharing

A fiber bundle is considered as a collection of N elastic fibers connected in parallel and with both ends clamped to some support. For simplicity, the fibers are assumed to be of equal length and to have the same elastic constant. The elongation x of the ith fiber is proportional to the force it carries up to a maximum value x_i. When the threshold value x_i is reached, the fiber fails and is no longer able to carry any force. Consequently, this fiber does not contribute to the bundle strength thereafter. Typically, not all fibers fail simultaneously due to statistical variations in fiber strength thresholds.

The individual threshold values x_i are assumed to be independent random variables with the same cumulative distribution function $P(x)$ and a corresponding density function $p(x)$:

$$\mathrm{Prob}(x_i \leq x) = P(x) = \int_0^x p(u)\ \mathrm{d}u. \tag{2.1}$$

There are no spatial correlation on how fiber strengths x_i are distributed over the bundle.

When a fiber fails, the load it carried must be taken up by the remaining intact fibers. The increased load on the surviving fibers may induce further failures, and the process will go on until all remaining fibers can withstand the load, or, alternatively, until the whole bundle breaks down.

Several mechanisms for redistribution of the load of a failed fiber on the surviving fibers have been used in the literature. The simplest one is to assume that the extra load is equally distributed on *all* the surviving fibers. This would be the case if the supports at the two clamped ends of the bundle are rigid. A bundle for which this democratic load sharing is obeyed is denoted an *equal-load-sharing model*. Equal load sharing is also called *global load sharing*. Daniels, in the first quantitative discussion of fiber bundles, assumed equal load sharing [9]. The equal load sharing applies to a *class* of models, differing in the distribution $P(x)$ of fiber strengths and by the bundle size N. Equal-load-sharing models are *mean-field models*, since every element is connected to every other element, and nothing depends on spatial separations.

The Fiber Bundle Model: Modeling Failure in Materials, First Edition.
Alex Hansen, Per C. Hemmer and Srutarshi Pradhan.

If one or both supports at the clamped bundle ends are *not* rigid, the surviving fibers in the neighborhood of a failed fiber will be most affected. Thereby short-range interactions are introduced, and such bundles are classified as *local-load-sharing models*. Several mechanisms for the load redistribution may be envisaged, and the ones in the literature fall into three groups. In the first group, it is assumed that only the *nearest-neighbor* surviving fibers share the extra stresses caused by a fiber failure. In the second group, the extra stress is taken up also by further neighbors with a fraction decreasing with increasing distance from the failed fiber. In both these groups, geometry, that is, how the fibers are arranged in a cross section of the bundle, is important. In the third group, it is assumed that at least one of the supports to which the bundle is attached is *soft* in the sense that it deforms as a continuous elastic medium when fibers are loaded. We defer the treatment of these local-load-sharing models to Chapter 4.

The class of equal-load-sharing models is the theme of this chapter and of Chapters 3, 5, and 6. Its relative simplicity often allows a more-or-less complete analytic treatment. In spite of its simplicity, this model class will have interesting and intriguing properties.

2.2
Strain-Controlled versus Force-Controlled Experiments

We consider a fiber bundle consisting of a large number N of parallel elastic fibers with no physical contact, to which a stretching force F is applied in the axial direction (Figure 2.1). The extension of the fibers is denoted by x, the same distance for all intact fibers. All fibers are assumed to obey linear elasticity right up to

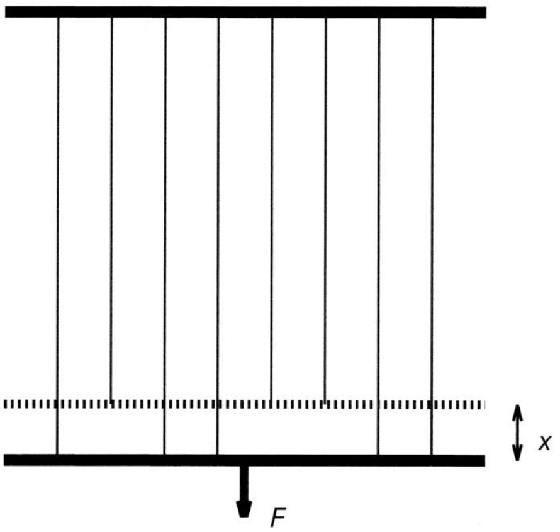

Figure 2.1 A fiber bundle model stressed by an external force F. A bundle is clamped between two rigid supports. The force has displaced one support a distance x from its original position (sketched). This has caused some fibers to fail, while other fibers are intact.

the breaking point. The fibers have the same elastic constant κ (which we, for simplicity, often normalize to unity), but different breaking thresholds x_i, picked randomly from a probability distribution $p(x)$.

The average total force on the bundle at an extension x is easy to assess. Since the probable number of fibers with thresholds x_i less than x is given by the cumulative probability distribution $P(x)$, the probable number of intact fibers at extension x equals $N(1 - P(x))$. The expected or *average force* $F(x)$ at extension x is, therefore,

$$F(x) = N\,\kappa\, x(1 - P(x)). \tag{2.2}$$

In the following, we set $\kappa = 1$, except when there are special reasons to have the elastic constant explicit. Here $F(0) = 0$ and since $F(x) = 0$ when all fibers have failed, the average force $F(x)$ is expected to go through a maximum. In the following, we will often refer to $F(x)$ as the *load curve*. In Chapter 3, where fluctuations enter the discussion, we shall, for clarity, write $\langle F \rangle$ instead of F, to make it explicitly clear that it is an average force.

A simple example is the case of thresholds uniformly distributed between 0 and 1, that is,

$$p(x) = \begin{cases} 1 & \text{for } 0 \le x \le 1 \\ 0 & \text{for } x > 1 \end{cases} \quad \text{or} \quad P(x) = \begin{cases} x & \text{for } 0 \le x \le 1 \\ 1 & \text{for } x > 1 \end{cases} \tag{2.3}$$

The average force in this case is a smooth parabolic function of x,

$$F(x) = Nx(1 - x). \tag{2.4}$$

However, due to the probabilistic nature of fiber failures, there will always be fluctuations around the average behavior. Therefore, the real force will be different from the average force. When the fiber bundle is stretched until the first fiber bursts, the elastic force will suddenly decrease at this value of x, with an amount equal to the elastic force carried by the failed fiber (Figure 2.2(a)). After the burst, the force again starts increasing linearly with the extension x. The proportionality constant is smaller, however, because the number of intact fibers is now less than N. Since the strain here is the independent variable, one is performing a *strain-controlled experiment.*

Experiments may be performed in a different way, however. In a *force-controlled experiment*, the applied external force F is increased slowly. The external force per fiber is the *stress*

$$\sigma = F/N. \tag{2.5}$$

When the first fiber fails in a force-controlled experiment, the bundle force F remains constant at the breaking point, while the extension x increases abruptly until the elastic forces of the intact fibers balance the applied force (Figure 2.2(b)). After the break, the extension starts increasing linearly with the force again.

Note that when one in Figure 2.2 redraws Figure (b) with x as abscissa, shown in (c), we obtain the same figure as in (a), if Figure (c) is supplemented with a

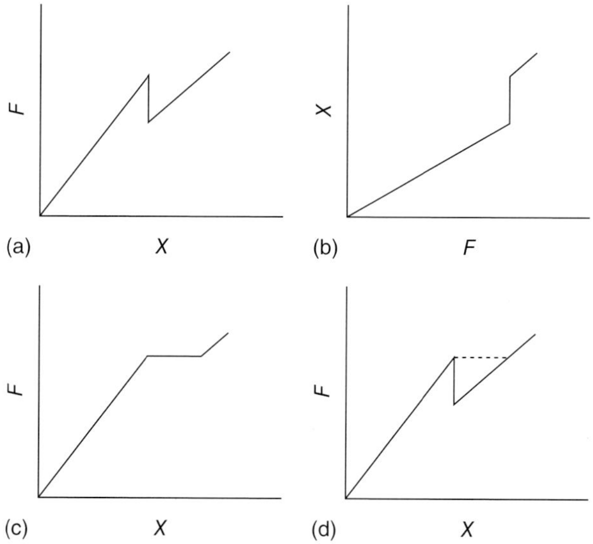

Figure 2.2 The behavior of the strain–force relation near the first fiber failure. (a) In a strain-controlled situation, the force drops abruptly. (b) In a force-controlled situation, the strain increases abruptly. Figure (c) is identical to (b) with coordinate axes interchanged. Figure (d) contains both the strain-controlled situation (fully drawn lines) and the force-controlled case with dashed lines.

horizontal line that makes F a monotonously increasing function of x. Thus, both the strain-controlled and the force-controlled situation may be represented in one figure, as in (d).

By a sudden increase in the extension x in a force-controlled experiment, thresholds of one or more fibers may be reached. Therefore, several fibers may fail simultaneously, without any increase in the external force. We will discuss in detail these interesting fluctuation-induced *burst avalanches.*

In Figure 2.3 is sketched an example of how the real elastic force may depend on the elongation x. In a strain-controlled experiment, the solid curve is followed. After a fiber fails in a force-controlled experiment, however, the system follows a horizontal constant-force line with increasing x until a new stable situation is reached or the whole bundle collapses. In the figure, this route is shown by the graph in which the decreasing parts of the load curve are replaced by horizontal dashed lines. The resulting graph can be characterized as the least monotonic function (LMF) of x that nowhere is less than the elastic force function $F(x)$.

The relevant function when the force is controlled is, therefore,

$$F_f(x) = \text{LMF}F(x), \tag{2.6}$$

where $\text{LMF}F(X)$ is the least monotonic function such that $\text{LMF}F(x) \geq F(x)$.

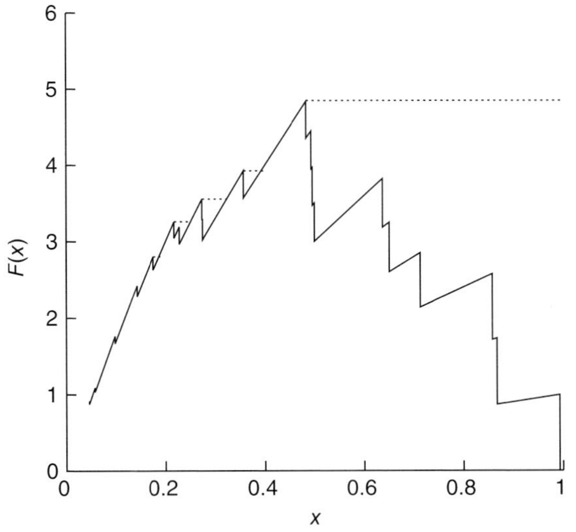

Figure 2.3 A sketch of how the real elastic force on the bundle may vary with increasing strain x for a finite N. In a strain-controlled experiment, the bundle follows the solid graph. In a force-controlled experiment, however, the system complies with the non-decreasing graph with the dashed lines.

Henry Daniels

Photo: Courtesy of the Statistical Laboratory, Cambridge, UK

Henry Ellis Daniels was born in London in 1912. His parents were Jewish and had fled from the pogroms in the Russian territories of Poland and Lithuania. His mother had arrived in Leith on a Scottish herring boat that had sold fish to the Baltic ports. The family moved to

Edinburgh, and Henry studied mathematics at the University of Edinburgh. He also spent 2 years at Cambridge University, where he specialized in statistics.

When in 1935, the Wool Industries Research Association in Leeds advertised for a mathematical statistician, Daniels applied and was hired. In addition to his theoretical activities on the statistical properties of yarns and fibers, he was also running a fiber-measuring laboratory. At Leeds, he did pioneering work on the strength of fiber bundles. Although the problem was stated in a textile context, his results were rather general. Daniels earned a PhD at the University of Edinburgh for his work on strength of bundles.

In 1957, Daniels was appointed Lecturer in Mathematics at the new Statistical Laboratory at Cambridge, and in 1957, he was hired as the first Professor of Mathematical Statistics at the University of Birmingham.

He was interested in a broad field in statistics. On a sabbatical at King's College, Cambridge, he organized a seminar on mathematical biology. He found diffusion-like approximations to Markov stochastic processes and studied boundary problems for Brownian motion. During 1974–1975, Daniels was President of the Royal Statistical Society.

Daniels was interested in music. A fine pianist, but later he concentrated on the English concertina (an instrument designed by the physicist Wheatstone). He was an expert on watches and helped design a watch showing sidereal as well as solar time, with an accuracy of half a second per year. In a lecture just before his death, he remarked that several results could be applied to mathematical finance, a topic he disliked: "soft pornography for the benefit of pure mathematicians," as he put it. Daniels died in 2000.

2.3 The Critical Strength

For the equal-load-sharing model described in Section 2.1, several exact results are available due to the mean-field nature of the system. Let us, for simplicity, assume that the bundle contains N parallel fibers with the same elastic constant (normalized to unity here). When the bundle is strained, the weakest fiber ruptures, and the process continues until all fibers fail. It is of interest to determine the maximum force $F_{\max}$ that the bundle can stand.

Let us start with a simple and somewhat unrealistic example in which the N threshold values x_i for breaking are *strictly uniformly* distributed,

$$x_i = x_{\mathrm{r}} i/N \quad \text{for } i = 1, 2, \ldots N. \tag{2.7}$$

Here x_{r} defines the length scale. For simplicity, we now put $x_{\mathrm{r}} = 1$. When the strain exceeds x_i, but not x_{i+1}, exactly i fibers are broken, and the total force on the bundle equals

$$F(x) = (N - i)x \quad \text{for } x_i \leq x < x_{i+1}. \tag{2.8}$$

The force per fiber, $F(x)/N$, is shown in Figure 2.4 for $N = 20$. When the elongation x (the strain) is the controlling parameter, the discontinuous solid curve is followed. In a force-controlled experiment, the decreasing parts of F are replaced by the constant values shown as dashed lines, as discussed in the preceding section.

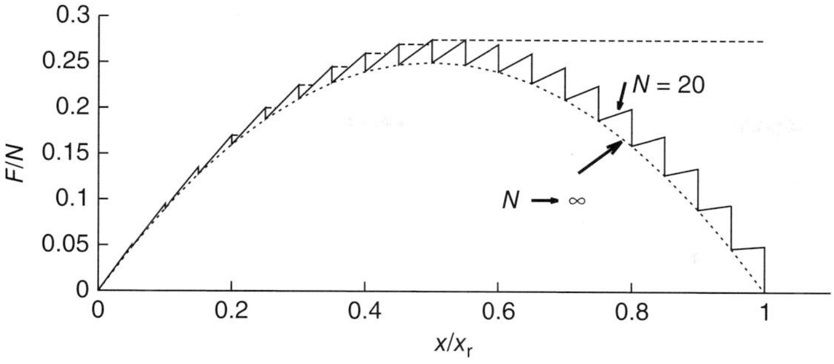

Figure 2.4 The solid curve represents for $N = 20$ the force per fiber, $F(x)/N$, as a function of x. The dashed lines show $F_f(x)$ when it exceeds $F(x)$. In the limit $N \to \infty$ the parabolic dotted curve is obtained.

When N is increased, the magnitudes of the discontinuities in F/N decrease and disappear in the limit $N \to \infty$. Since $i/N = x_i$, we have $F(x_i)/N = (1 - x_i)x_i$. The limiting curve is, therefore,

$$F(x)/N \;\to\; x(1-x) \quad \text{for } N \to \infty. \tag{2.9}$$

The maximum of the parabola, occurring at $x = \frac{1}{2}$, equals the maximum force the bundle may tolerate. We denote such a maximum of the load curve as the *critical force* F_c. In this case,

$$F_c/N = \frac{1}{4} \quad \text{for large } N, \tag{2.10}$$

in our dimensionless units.

Note that for N finite, the maximum strength per fiber exceeds the maximum F_c of the average force. This will also be seen in more general settings (Section 3.2).

We will, in general, be interested in systems in which we do not know the precise fiber strengths (thresholds), but merely their probability distribution. As mentioned earlier, the threshold values x_i are picked from a continuous cumulative distribution $P(x)$ with a probability density $p(x)$:

$$\text{Prob}(x_i \le x) = P(x) = \int_0^x p(u)\, du. \tag{2.11}$$

At an elongation x, $N(1 - P(x))$ number of fibers are expected to be intact, so that the average total force on the bundle in a strain-controlled experiment will be

$$F(x) = N\, x[1 - P(x)], \tag{2.12}$$

Eq. (2.2). It is convenient to introduce the force per fiber, the stress

$$\sigma(x) = F/N = x[1 - P(x)]. \tag{2.13}$$

The maximum force in this macroscopic treatment (large N) appears at the value $x = x_c$ for which dF/dx vanishes. This gives

$$1 - P(x_c) - x_c p(x_c) = 0. \tag{2.14}$$

We denote x_c as *the critical point*. The maximum total force per fiber is

$$\sigma_c = x_c\,[1 - P(x_c)] \quad \text{or} \quad \sigma_c = x_c^2\, p(x_c), \tag{2.15}$$

since the critical stress is equal to the maximum force per fiber. Formula (2.14) is used. In a force-controlled experiment on a fiber bundle, the critical point signals where the complete breakdown of the whole bundle takes place.

Note, however, that for finite N, the real maximum force per fiber is not equal to the maximum σ_c of the average force. The difference is due to fluctuations around the average force, and the fluctuations are small when N is large. This situation was seen clearly in the exactly solvable example with a strictly uniform fiber strength distribution presented (Figure 2.4).

Two popular examples of threshold distributions are the uniform distribution

$$P(x) = \begin{cases} x/x_r & \text{for } 0 \le x \le x_r \\ 1 & \text{for } x > x_r, \end{cases} \tag{2.16}$$

and the Weibull distribution

$$P(x) = 1 - \exp\,[-(x/x_r)^k] \quad \text{for } x \ge 0. \tag{2.17}$$

Here x_r is a reference threshold and the dimensionless shape parameter k is the Weibull index (Figure 2.5).

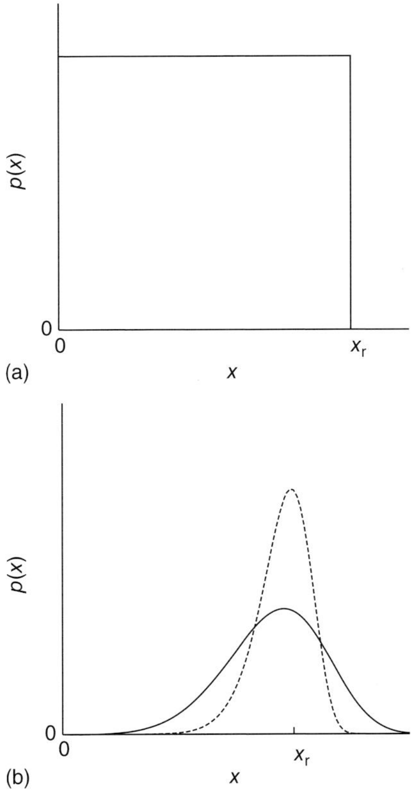

Figure 2.5 The uniform distribution (a) and the Weibull distribution (b) with $k = 5$ (solid line) and $k = 10$ (dotted line).

The uniform distribution is often used because its simplicity frequently makes it possible to obtain many model results analytically, as we will see in the following.

The Weibull distribution, named after Waloddi Weibull, is extensively used in reliability engineering and elsewhere, due to its versatility and relative simplicity [17]. It is also one of the asymptotic distributions in extreme value statistics (see the Statistical Toolbox, Appendix B).

Waloddi Weibull

Photo credit Prof. Sam C. Saunders

Ernst Hjalmar Waloddi Weibull was born in 1887 in Sweden. His family had emigrated from Schleswig-Holstein, then Danish, in the eighteenth century. Weibull started his career as a navy officer. He entered the Swedish Coast Guard in 1904 as a midshipman and moved up the ranks: sublieutenant in 1907, captain in 1916, and major in 1940. At the same time, he took courses at the Royal Institute of Technology (KTH), Stockholm. At KTH, he received, in 1924, the licentiate degree and was appointed Professor of Machine Elements the same year. During 1931 – 1940, he was Prorector at the institution. He received an honorary doctorate at Uppsala University in 1932.

Weibull was a dynamic and accomplished engineer. His scientific interests were very varied and always with a practical objective. His first paper, in 1914, was on sea seismics, on the method of using reflected waves from explosions to research the type of sea bed sediments and their thickness. He worked for Swedish and German industries as a consultant and as an inventor: ball bearings and the electric hammer are examples. In 1941, Bofors, a Swedish arms factory, financed a personal research professorship in Technical Physics at KTH for him. Weibull's main interests centered around strength of materials, fatigue, reliability, and rupture in solids. He authored in 1961 a book Fatigue Testing and Analysis of Results. His most well-known work is on the Weibull distribution in statistics. It appeared first in 1939 and then in his most famous paper presented in 1951 to the American Society of Mechanical Engineers [17]. He demonstrated through case studies the usefulness of allowing the data select the most appropriate distribution from the broad class of Weibull distributions.

Weibull did not derive his distribution from a theoretical basis. Later it turned out, however, that the Weibull distribution is one of the three limiting distributions in extreme statistics (see Appendix B.2). Weibull received many honors for his achievements. He died in Annecy, France, at the age of 92.

For the uniform distribution with $x_r = 1$, Eq. (2.13) yields

$$\sigma(x) = x(1 - x). \tag{2.18}$$

Hence, the critical strength per fiber $\sigma_{max} = \frac{1}{4}$ occurs at $x_c = \frac{1}{2}$. Note that the *statistically uniform* distribution (2.16) is not the same as the *strictly uniform* distribution (2.7) considered, although their *average* properties are identical in the large N limit.

For the Weibull distribution with $x_r = 1$, the average force per fiber (2.13) takes the form

$$\sigma(x) = x \exp(-x^k), \tag{2.19}$$

with a parabolic maximum at

$$x_c = k^{-1/k}. \tag{2.20}$$

Hence, the critical strength per fiber, σ_c, takes the form

$$\sigma_c = (ke)^{-1/k}. \tag{2.21}$$

Figure 2.6 shows how the critical strength depends on the Weibull index k. It has a minimum at $k = 1$ and increases slowly with increasing k.

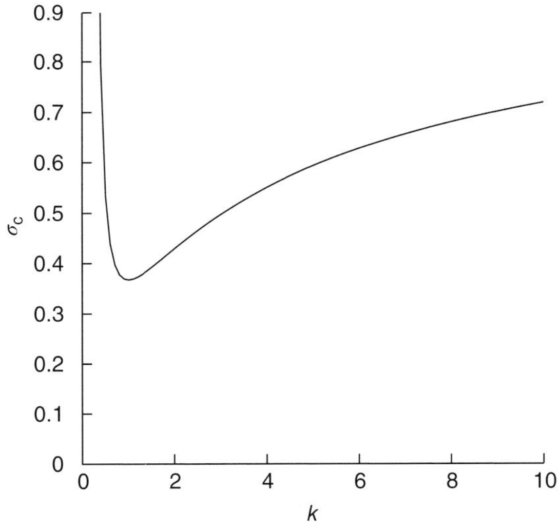

Figure 2.6 The critical strength per fiber, σ_c, for a fiber bundle with thresholds satisfying the Weibull distribution (2.16), as a function of the Weibull index k.

For most reasonable threshold distributions, the load curve $F(x)$ will have a unique parabolic maximum at the critical point x_c. The critical value $F_c = F(x_c)$ will be the maximum average force the fiber bundle can withstand. This is the generic situation. However, exceptions to this generic situation exist, and we give now some examples of such exceptional cases [18].

Firstly, a threshold distribution may give rise to *several parabolic maxima* in the force. For a simple example, consider the following piecemeal uniform distribution:

$$p(x) = \begin{cases} p_1 & \text{for } 0 \leq x < a \\ p_2 & \text{for } a \leq x \leq 1. \end{cases} \tag{2.22}$$

Normalization requires $a = (1 - p_2)/(p_1 - p_2)$. The corresponding average force (2.13) takes the form

$$\sigma(x) = \begin{cases} x - p_1 x^2 & \text{for } 0 \leq x < a \\ p_2(x - x^2) & \text{for } a \leq x \leq 1. \end{cases} \tag{2.23}$$

Figure 2.7(a) shows that for the values $p_1 = 4$ and $p_2 = 1/3$, the force function has *two* parabolic maxima. For $x < a$, the strong fibers are merely unruffled spectators to the breaking process of the weak fibers, and for $x > a$, all the weak fibers are already eliminated, while the breaking of the strong fibers proceeds. Each rupture process gives, apart from scaling, a parabolic maximum equal to (2.18) for the uniform distribution. The maximum strength $\sigma_c = \frac{1}{12}$ is determined by the highest extremum at $x = \frac{1}{2}$.

Secondly, the force function $\sigma(x)$ may have a maximum that is *not* parabolic. For an interesting example, take the threshold distribution (2.22) with the values $p_1 = 0$ and $p_2 = 3$. The corresponding force function is shown in Figure 2.7(b). In this case, no fibers break as long as the applied force is weaker than the maximum value σ_{max} of $\sigma(x)$. The maximum strength per fiber is in this case $\sigma_{max} = a = 2/3$. However, when σ_{max} is reached, all fibers break simultaneously, as long as fluctuations are ignored. Such instant failures require the presence of a low cut-off in the threshold distribution, and, moreover, the average force takes its maximum value at the cut-off point.

Thirdly, the force function $F(x)$ may be steadily increasing with *no maximum* at all. As a simple example, consider the threshold distribution

$$P(x) = \begin{cases} 0 & \text{for } 0 \leq x \leq 1 \\ 1 - 1/\sqrt{x} & \text{for } x > 1 \end{cases} \tag{2.24}$$

By (2.13) the average force,

$$\sigma(x) = \begin{cases} x & \text{for } 0 \leq x \leq 1 \\ \sqrt{x} & \text{for } x > 1 \end{cases} \tag{2.25}$$

is increasing forever (see Figure 2.8). Hence, the fiber bundle never fails completely. This is due to the never-ending supply of fibers of sufficient strength. A manifestation of this is that the threshold distribution (2.25) has a *divergent* average fiber strength $\langle x \rangle$.

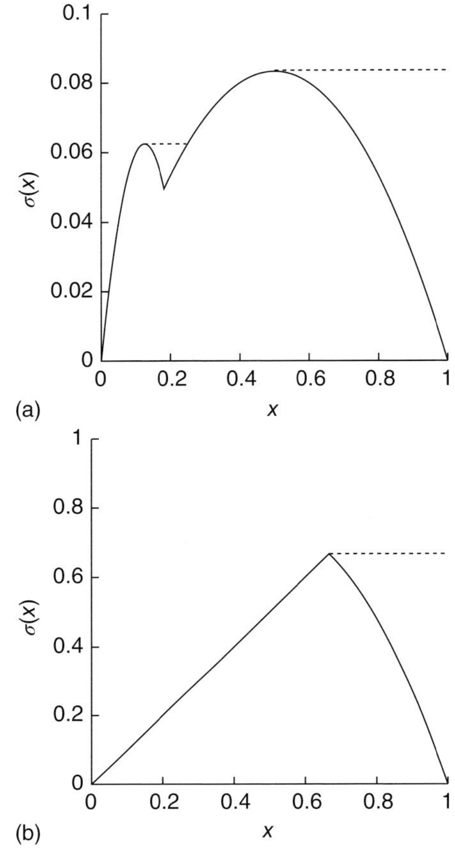

Figure 2.7 The average force per fiber,$\sigma(x)$, as a function of x for the piecemeal uniform distribution (2.22) with (a) $p_1 = 4, p_2 = 1/3$, and (b) $p_1 = 0, p_2 = 3$.

2.4 Fiber Mixtures

It is in principle easy to extend the treatment to bundles consisting of a mixture of fibers with different elastic constants κ_i and different probability distribution functions $P_i(x)$ for fiber strengths [19]. The total average force is the sum of the contributions from each fiber type. With N_i fibers of type i, the average force at extension x equals

$$F(x) = \sum_i N_i \kappa_i x[1 - P_i(x)], \tag{2.26}$$

a straightforward generalization of (2.2).

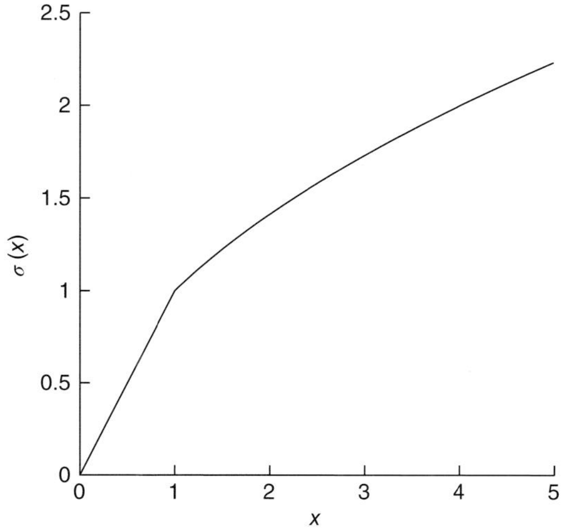

Figure 2.8 The average force per fiber, $\sigma(x)$, as a function of x for the threshold distribution (2.25).

With a continuous distribution of elastic constants, so that there are $N_\kappa q(\kappa)\ \mathrm{d}\kappa$ fibers with constants in $(\kappa, \kappa + \mathrm{d}\kappa)$, the discrete version (2.26) must be replaced by

$$F(x) = \int N_\kappa\ \kappa\, x\, [1 - P_\kappa(x)]\, g(\kappa)\ \mathrm{d}\kappa, \tag{2.27}$$

where $P_\kappa(x)$ is the distribution of thresholds for fibers with elasticity constant κ.

Here we merely give a simple example. Consider two fiber bundles 1 and 2, so that bundle n consists of fibers with elastic constant κ_n, and with a uniform threshold distribution, $P_i(x) = x/x_n$ for $x \leq x_n$ and equal to 1 for $x > x_n$. The maximum average strength per fiber, σ^0_{max} for fiber bundle 1 equals the maximum of the force per fiber, $\sigma(x) = \kappa_1 x(1 - x/x_1)$, that is,

$$\sigma^0_{\mathrm{max}} = \frac{1}{4}\kappa_1 x_1. \tag{2.28}$$

Assume that bundle 2 has the same maximum strength per fiber as bundle 1, which requires $x_2 = x_1\kappa_1/\kappa_2$. Consider now a mixed bundle, consisting of $N/2$ fibers from bundle 1 and $N/2$ fibers from bundle 2. In this case, the average force (2.26) equals

$$F(x) = \frac{1}{2}N \sum_{n=1}^{2} \kappa_n\ x(1 - x/x_n), \tag{2.29}$$

for x smaller than x_1 and x_2.

The maximum strength occurs at an extension x_c satisfying

$$F'(x_c) = \frac{1}{2}N \sum_{n=1}^{2} \kappa_n(1 - 2x_c/x_n) = 0. \tag{2.30}$$

Solving for the critical value x_c and inserting into (2.29), we find the critical strength

$$F_c = \frac{1}{8}N\frac{(\kappa_1+\kappa_2)^2}{\kappa_1/x_1+\kappa_2/x_2} = \frac{1}{8}N\kappa_1 x_1\frac{(\kappa_1+\kappa_2)^2}{\kappa_1^2+\kappa_2^2}, \tag{2.31}$$

when $x_2 = x_1\kappa_1/\kappa_2$ is inserted. Comparison with the result (2.28) gives the maximum strength per fiber for the mixed bundle equal to

$$\sigma_c = \sigma_c^0\frac{(\kappa_1+\kappa_2)^2}{2(\kappa_1^2+\kappa_2^2)} = \sigma_{\mathrm{max}}^0\left[1-\frac{(\kappa_1-\kappa_2)^2}{2(\kappa_1^2+\kappa_2^2)}\right] \tag{2.32}$$

Thus, the mixed fiber bundle is weaker than each of the single-component bundles from which it is put together. The mixed bundle has the same maximal strength as the single-component bundles *only* when the two types of fibers have identical elastic properties.

2.5 Non-Hookean Forces

So far, it has been assumed that Hooke's law holds for fibers, so that the force F is strictly proportional to the extension x up to the breaking point. Non-Hookean forces can take many forms; here we merely give two very simple examples.

2.5.1 Fibers with Random Slacks

Let fiber n have a slack ξ_n so that the force on the fiber at extension x equals $\kappa(x - \xi_n)$ for $\xi_n \leq x \leq \xi_n + x_n$ and is zero otherwise (Figure 2.9) [19]. The N slacks ξ_n are assumed to be distributed independent of each other according to a density

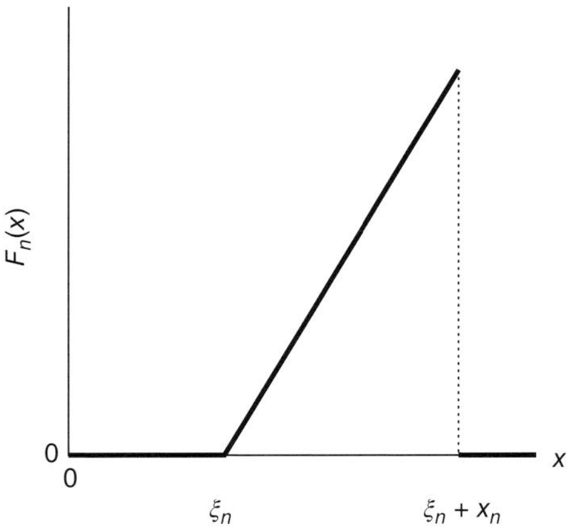

Figure 2.9 The force on fiber n at extension x. The size of the slack is ξ_n.

function $g(\xi)$ and uncorrelated with the fiber strengths x_n. The fiber strengths x_n are distributed according to a cumulative probability function $P(x-\xi_n)$, which vanishes for $x < \xi_n$.

The average force on the bundle at extension x is then

$$\langle F(x)\rangle = N\kappa \int_0^x (x-\xi)[1-P(x-\xi)]\, g(\xi)\, \mathrm{d}\xi. \tag{2.33}$$

As an example, take a uniform distribution of fiber strengths on the unit interval, $P(x) = x$ for $x \leq 1$. We find

$$\langle F(x)\rangle = N\kappa\langle x-\xi-(x-\xi)^2\rangle = N\kappa[1/4-\sigma_\xi^2-(x-1/2-\langle\xi\rangle)^2], \tag{2.34}$$

where $\sigma_\xi^2 = \langle(\xi-\langle\xi\rangle)^2\rangle$ is the variance of the slack distribution. The maximum macroscopic force is, therefore,

$$\langle F\rangle_{\mathrm{c}} = N\kappa\left(1/4-\sigma_\xi^2\right), \tag{2.35}$$

and occurs at an extension

$$x_{\mathrm{c}} = 1/2+\langle\xi\rangle. \tag{2.36}$$

Thus, the presence of slacks *lowers* the maximum average force (except in the trivial case when all the slacks have the same size). The critical point x_{c} in the presence of slacks is shifted with an amount equal to the average slack $\langle\xi\rangle$, as could be expected.

Let us more generally show that

$$\langle F\rangle_{\mathrm{c}}\ \text{(with slacks)} \leq \langle F\rangle_{\mathrm{c}}\ \text{(without slacks)} \tag{2.37}$$

for any fiber strength distribution and for any slack distribution.

As a generalization of the fact that the maximum of a sum of positive terms is less or equal to the sum of the maxima of the separate terms, Eq. (2.33) gives the inequality

$$\langle F\rangle_{\mathrm{c}} \leq \int \mathrm{d}\xi\, g(\xi)\, \{N\kappa(x-\xi)[1-P(x-\xi)]\}_{\mathrm{c}}. \tag{2.38}$$

Since the maximum of $N\kappa(x-\xi)[1-P(x-\xi)]$ is independent of ξ, and equal to $\langle F\rangle_{\mathrm{c}}$ (without slacks), the integral boils down to this factor times $\int g(\xi)\,\mathrm{d}\xi = 1$. Hence, the inequality (2.37) follows.

2.5.2
Elastic–Plastic Model

Let us consider fibers i that elongate according to Hooke's law up to elongations x_i and then continue to be stretched plastically at constant load until failure at $x = \hat{x}_i$ (Figure 2.10) [19].

The threshold values that mark the end of the elastic regime follow a probability distribution $P(x)$, with corresponding density $p(x)$, as before. The plastic elongations $\hat{x}_i$ are apportioned according to another distribution $\hat{P}(\hat{x})$. For simplicity, we assume that the two types of elongations are uncorrelated.

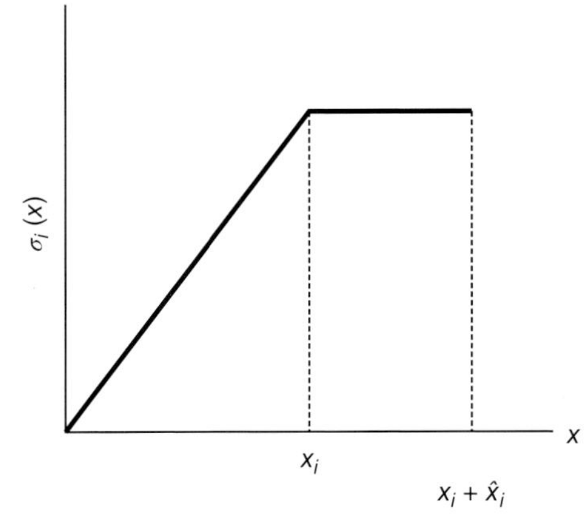

Figure 2.10 The force per fiber, σ_i, on fiber i as function of its elongation x. The elastic regime is $x \in (0, x_i)$, the plastic regime corresponds to $x \in (x_i, x_i + \hat{x}_i)$. The constant force in the plastic domain equals $\kappa\ x_i$.

At an elongation corresponding to the threshold value x, the force contributions due to the plastic deformations come from fibers with lower thresholds, $x' < x$. For fibers with thresholds in an interval $(x', x' + \mathrm{d}x')$, the plastic force contribution from each fiber is $\kappa x'$, and the average number of fibers contributing to the force at x equals $N[1 - \hat{P}(x - x')]$. Thus, the average force expression (2.12) generalizes in this case to

$$\langle F(x) \rangle = N\kappa x[1 - P(x)] + N \int_0^x p(x')\kappa x'[1 - \hat{P}(x - x')]\, \mathrm{d}x'. \tag{2.39}$$

Clearly, the maximum of the average force is now *higher* than it would have been without the plastic elongations.

In this chapter, we have only considered average properties of fiber bundle models. Now we turn to the much more interesting features that are caused by *fluctuations* around the average behavior.

3
Fluctuation Effects

In Chapter 2, we studied *average* values for bundles with a very large number of fibers and with equal load sharing. However, since the strengths of individual fibers are statistically distributed, there will be *fluctuations* around such average values. And the fluctuations do not merely give small additions to the average values. They are essential for the most interesting features, the breakdown properties in fiber bundle dynamics. Not only for the fiber bundle model, but in material science in general, fluctuations are a main ingredient in fracture and breakdown phenomena.

We start by investigating in Section 3.1 the range of force fluctuations. Fluctuation effects are connected with a bundle consisting of a *finite* number N of fibers. In the beginning of Chapter 2.3, we saw by the example with a strictly uniform distribution of fiber strengths how the fluctuations in the force per fiber, F/N, disappear in the $N \to \infty$ limit. This example is also noteworthy in that these deviations from the average load curve exist even without any statistical distribution of the individual fiber strengths. In the presence of a statistical distribution of fiber strengths, the deviations from the average behavior will increase, as we shall see.

In Section 3.2, we investigate briefly how fluctuations affect the maximum bundle strength. Interesting scaling behavior with N, the number of fibers in the bundle, is seen.

Section 3.3 is devoted to a quantitative study of *rupture burst*. When a fiber fails, the stress on the intact fibers increases. This may in turn trigger further fiber failures and so on. The size Δ of a burst is defined as the number of fibers that rupture simultaneously, without complete bundle failure. We determine, in particular, the size distribution of bursts, the expected number $D(\Delta)$ of bursts of a given size Δ that occur during the entire breakdown process until complete bundle failure.

In this chapter, we also determine the distribution of mechanical energy released in bursts. Finally, we discuss the distribution of burst sizes when the load on the bundle is not increased continuously, but in finite steps.

The Fiber Bundle Model: Modeling Failure in Materials, First Edition.
Alex Hansen, Per C. Hemmer and Srutarshi Pradhan.

3.1 Range of Force Fluctuations

Examples of force fluctuations can be seen in Figure 3.1. With our dimensionless elastic constant, $\kappa = 1$, , the elastic force $F(x) = N\sigma(x)$ on the bundle with N fibers at extension x equals the extension times the number s of intact fibers:

$$\sigma(x) = x\ s(x)/N. \tag{3.1}$$

The number $s(x)$ of intact fibers is the number of fibers with thresholds larger than x. If N threshold values are distributed independently according to a cumulative probability distribution $P(x)$, the probability $q_x(s)$ that s threshold values are larger than x equals

$$q_x(s) = \binom{N}{s} [1 - P(x)]^s\ [P(x)]^{N-s}. \tag{3.2}$$

It is straightforward to evaluate the average $\langle s \rangle = N[1 - P(x)]$ and the variance $\langle (s - \langle s \rangle)^2 \rangle = N\,P(x)[1 - P(x)]$. For the force per fiber, Eq. (3.1), this gives the average

$$\langle \sigma \rangle = x\ [1 - P(x)], \tag{3.3}$$

as is already known from Chapter 2, relation (2.13).

The variance of the force at given extension x equals

$$\langle (\sigma - \langle \sigma \rangle)^2 \rangle = x^2 P(x)\ [1 - P(x)]/N. \tag{3.4}$$

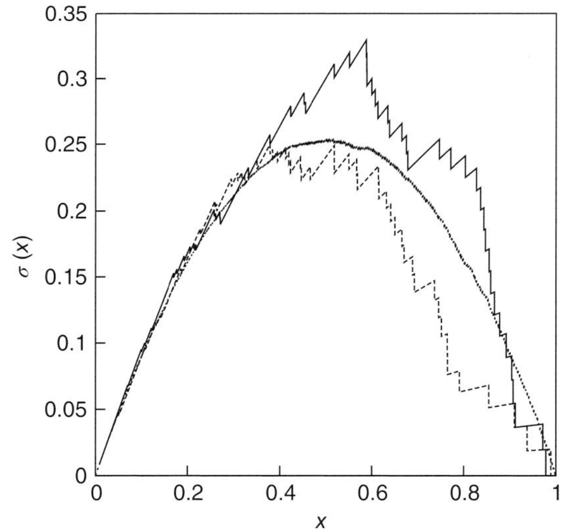

Figure 3.1 Two realizations for the force per fiber $\sigma(x)$ for a bundle with $N = 50$ fibers, as a function of the extension x. The uniform fiber strength distribution is assumed. For comparison, a realization with $N = 5000$ is shown. For such a large number of fibers, fluctuations are tiny, so that the resulting force per fiber deviates little from the parabolic average force, $\langle \sigma(x) \rangle = x(1 - x)$, for this model.

Thus, there is around the average force a layer of the size of the standard deviation

$$\delta_\sigma(x) \propto N^{-1/2} \tag{3.5}$$

in which most of the force fluctuations occur.

For large N and s, the distribution (3.2) is close to a Gaussian distribution with the average and variance given above. For σ, the force per fiber, this corresponds to a probability distribution

$$Q_x(\sigma) = \left(\frac{N}{2\pi x^2 P(x)[1-P(x)]}\right)^{1/2} \exp\left[-\frac{N(\sigma - \langle\sigma\rangle)^2}{x^2 P(x)[1-P(x)]}\right]. \tag{3.6}$$

One may consider a different kind of fluctuations, namely the fluctuations in the force precisely when the kth fiber is about to fail. At this moment, the force is the number of intact fibers times the extension:

$$\sigma_k = (N+1-k)x_k/N. \tag{3.7}$$

Thus, the average and the variance of the force take the form

$$\langle\sigma_k\rangle = (N+1-k)\langle x_k\rangle/N, \tag{3.8}$$

and

$$\langle(\sigma_k - \langle\sigma_k\rangle)^2\rangle = (N+1-k)^2 N^{-2}(\langle x_k^2\rangle - \langle x_k\rangle^2). \tag{3.9}$$

For simplicity, we evaluate the averages for a bundle with a uniform distribution of fiber strengths, $p(x) = 1$ for $0 \le x \le 1$. The relevant probability density follows from the probability $q_k\ \mathrm{d}x$ to find a threshold in the interval $\mathrm{d}x$ at x, with $k-1$ threshold values below x and $N-k$ threshold values above x:

$$q_k(x)\,\mathrm{d}x = \mathrm{d}x\,\frac{N!}{(k-1)!(N-k)!}\,x^{k-1}(1-x)^{N-k}. \tag{3.10}$$

The averages are easy to compute:

$$\langle x_k\rangle = \frac{k}{N+1} \quad \text{and} \quad \langle x_k^2\rangle = \frac{k(k+1)}{(N+1)(N+2)}, \tag{3.11}$$

which gives the variance of the force per fiber equal to

$$\langle(\sigma_k - \langle\sigma_k\rangle)^2\rangle = \frac{k(N+1-k)^3}{N^2(N+1)^2(N+2)} \simeq \langle x_k\rangle(1-\langle x_k\rangle)^3/N \tag{3.12}$$

for large N.

This variance is different from the variance (3.4) at constant extension, evaluated at the extension where the kth fiber is expected to break, $x = \langle x_k\rangle$. It is not surprising that they are different, since the variance at this value of x will represent bundles with many different numbers of intact fibers. However, also in this case, the standard deviation of the force per fiber is proportional to $N^{-1/2}$.

3.2 The Maximum Bundle Strength

As we have seen, the *average* (or macroscopic) bundle strength $\langle F\rangle$ has its maximum F_c at the critical point x_c. Due to fluctuations around this average value, the bundle strength is expected to reach, somewhere in the neighborhood of the critical point, a maximum that exceeds F_c . The average strength is proportional to N number of fibers, and it is of interest to determine how the expected maximum value F_{max} of the bundle strength scales with N,

$$F_{max} - F_c \propto N^{\gamma}. \tag{3.13}$$

In the generic case, the average force has a parabolic maximum at the critical point. Near criticality $\langle F\rangle$ is of the form

$$\langle F\rangle = F_c - N\,a\,(x - x_c)^2. \tag{3.14}$$

The constant a depends on the threshold distribution $P(x)$ and is determined by the load curve (2.12), $\langle F\rangle = Nx[1 - P(x)]$, as

$$a = -\frac{1}{2}\langle F\rangle''(x_c)/N = 2p(x_c) + x_c\,p'(x_c). \tag{3.15}$$

The force fluctuations seen in Section 3.1 were to be of order $N^{1/2}$. We assume that the fluctuations may be considered as standard Brownian motion $W(x)$, the continuum limit of a discrete random walk (see Appendix A.5). The Brownian motion has, for all t, zero average, and from a starting point $W(0) = 0$, the variance increases linearly:

$$\langle W(x)\rangle = 0, \qquad \langle W(x)^2\rangle = x. \tag{3.16}$$

Upon scaling t with a factor c, the Brownian motion scales as

$$c^{-1/2}\,W(cx) = W(x). \tag{3.17}$$

This is true because both sides of this equation have the same zero average and by (3.16) the same variance.

In the present problem, we are looking for the maximum of the total force

$$F = F_c - N\,a\,(x - x_c)^2 + N^{1/2}W(x). \tag{3.18}$$

Let us scale x,

$$x = N^{-\alpha}\xi, \tag{3.19}$$

so that the two terms in $F - F_c$ balance each other. The scaling property (3.17) of Brownian motion gives in the critical domain [20]

$$F - F_c = -N^{1-2\alpha}(\xi - \xi_c)^2 + N^{(1-\alpha)/2}\,W(\xi). \tag{3.20}$$

By taking $\alpha = 1/3$, both terms are of the same order in N:

$$F - F_c = N^{1/3}\,[W(\xi) - a(\xi - \xi_c)^2]. \tag{3.21}$$

So, the average value of the maximum strength is of the form

$$\langle F_{\max} \rangle = F_c + K\,N^{1/3}, \tag{3.22}$$

where

$$K = \langle m \rangle \qquad \text{with} \qquad m = \max_{\xi}\,[W(\xi) - a(\xi - \xi_c)^2]. \tag{3.23}$$

On a force per fiber basis, Eq. (3.22) corresponds to

$$\langle \sigma_{\max} \rangle = \sigma_c + K\,N^{-2/3}. \tag{3.24}$$

To check the scaling, we have performed a simulation study of the average maximum force for bundles with a uniform fiber strength distribution on (0,1), for bundles containing different number, N, of fibers. The evidence, shown in Figure 3.2, indicates clearly that $\langle \sigma_{\max} \rangle - \sigma_c$ scales as $N^{-2/3}$.

To determine the constant K, Eq. (3.23), it is necessary to carry through a highly nontrivial Brownian process. To dominating order in N, the final result is [21]

$$K = \lambda\,\sigma_c^{2/3}\,a^{-1/3} = 0.99615\,\sigma_c^{2/3}\,a^{-1/3}, \tag{3.25}$$

where $a = 2p(x_c) + x_c p'(x_c)$ by Eq. (3.15). The constant $\lambda = 0.99615$ is expressed by an integral over Airy functions. For thresholds distributed uniformly on (0,1), this yields

$$K = \lambda\,2^{-5/3} = 0.3138, \tag{3.26}$$

fully consistent with the simulations shown in Figure 3.2.

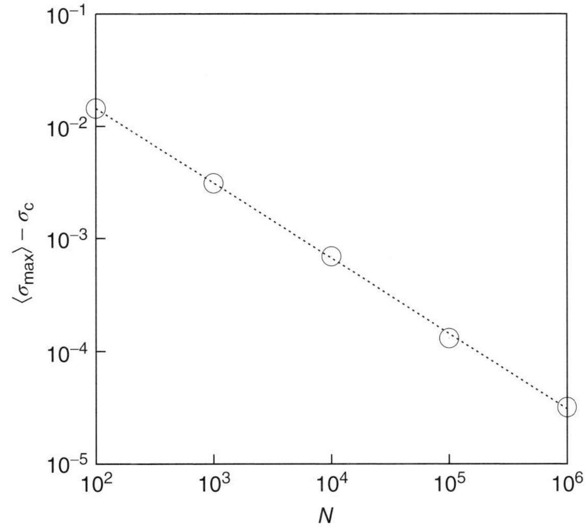

Figure 3.2 The average bundle strength $\langle \sigma_{\max} \rangle$ as function of the number N of fibers, for bundles with a uniform distribution of the fiber strengths. The dotted line represents a power law with an exponent of −2/3. The simulations are based on 10 000 samples for each value of N.

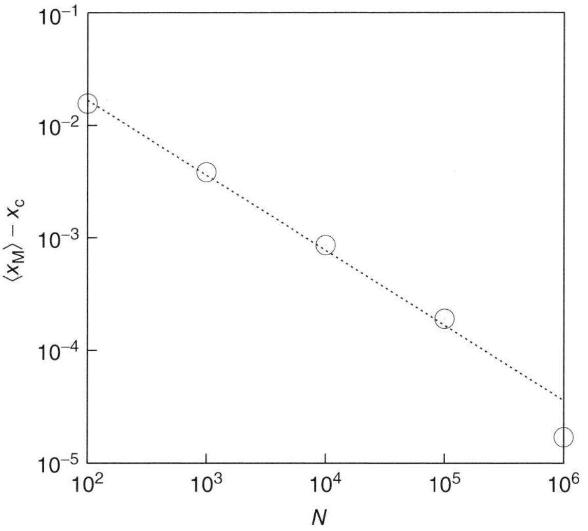

Figure 3.3 The average extension beyond criticality, $\langle x_M \rangle - x_c$, at which the maximum force occurs, as function of the number of fibers N. The bundles are assumed to have a uniform distribution of the fiber strengths. The dotted line represents the power law (3.28). The simulations are based on 10 000 samples for each value of N.

Is the maximum force most likely to occur at the critical extension x_c? While the average force is symmetric around x_c near criticality, the force fluctuations (3.4) are not. It is straightforward to show that the variance of the force, δ_σ^2, has a positive derivative at the critical point:

$$\left[\frac{\partial \delta_\sigma^2}{\partial x}\right]_c = \sigma_c/N > 0. \tag{3.27}$$

Since the fluctuations are larger for $x > x_c$ than for $x < x_c$, one expects that on average the maximum bundle force appears at an extension $x_M > x_c$. A simulation, again for the uniform fiber strength distribution, shows that this is the case (see Figure 3.3) and that the average of $x_M - x_c$ also has a power-law scaling,

$$\langle x_M \rangle - x_c \simeq d\, N^{-2/3}, \tag{3.28}$$

with a constant $d \simeq 0.33$ for this threshold distribution.

3.3 Avalanches

The most interesting effect produced by the presence of fluctuations is the existence of *burst avalanches*. An avalanche is an event in which several fibers fail simultaneously. The cause of such simultaneous failures is that when a fiber fails, the load carried by this fiber is, by the equal-load-sharing assumption, redistributed on the remaining intact fibers, and this increased load may induce further failures. The failure process continues until the remaining fibers can withstand the load or, alternatively, until the complete bundle breaks down.

We will study the rupture process under a steadily increasing external load on a bundle consisting of a large number N of fibers. Our main aim is to determine the expected distribution $D(\Delta)$ of the total number of avalanches of a given size Δ, when the bundle is loaded until complete breakdown. The most important outcome [18, 22] of the analysis is that in the generic case, the asymptotic distribution is given by a power law,

$$D(\Delta)/N \propto \Delta^{-5/2}, \tag{3.29}$$

with a universal exponent 5/2. This will now be shown.

3.3.1 The Burst Distribution

Let us order the fiber thresholds x_k sequentially according to magnitude, with fiber number 1 the weakest one. When the kth fiber is about to burst, the total force on the bundle is, therefore, equal to the strength of this fiber times the number of surviving fibers,

$$F_k = (N + 1 - k)x_k. \tag{3.30}$$

Since the thresholds x_k are statistically distributed, the forces F_k do not vary monotonically with k, but will fluctuate in relative sizes. These fluctuations may lead to bursts in which several fibers break simultaneously. An example is shown in Figure 3.4.

Consider now a burst of size Δ that is initiated by failure of fiber number k with threshold $x_k = x$. This requires two conditions on the forces F_j to be fulfilled. The

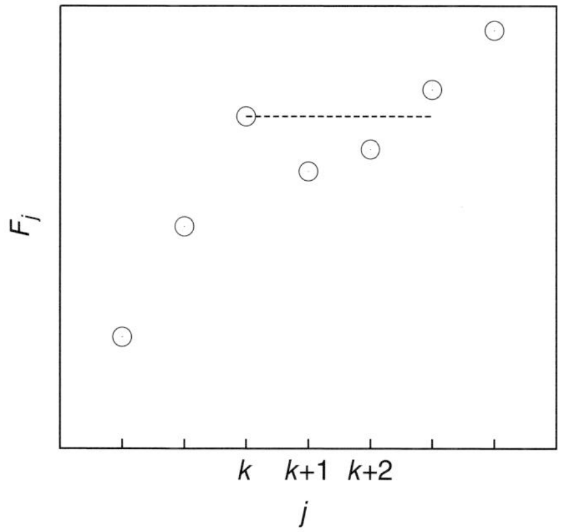

Figure 3.4 The figure shows an example of how the sequence F_j may vary with the fiber number j. When the external load compels fiber k to fail, the fibers $k+1$ and $k+2$ must necessarily also rupture at the same time. Thus, a burst of size $\Delta = 3$ will take place in this example.

first requirement is *the forward condition* that precisely $\Delta - 1$ additional fibers fail, that is,

$$F_{k+j} < F_k \quad \text{for } j = 1, 2, \ldots, \Delta - 1, \quad \text{while } F_{k+\Delta} > F_k. \tag{3.31}$$

The second requirement is *the backward condition*

$$F_j < F_k \quad \text{for all } j < k. \tag{3.32}$$

The backward condition (3.32) secures that the burst, starting with failure of fiber number k, is not merely an internal part of a larger burst.

By Eq. (3.30), the conditions on the F_j are conditions on the threshold values x_j. In a burst of a finite size Δ, the threshold values of the failed fibers cover just a tiny part of the range of all the N thresholds, and therefore, all threshold values of interest in the burst are close to the initial value x_k. The threshold value $x_{k+\Delta-1}$ of the strongest fiber in the burst can be estimated to be

$$x_{k+\Delta-1} \simeq x_k + \frac{\Delta}{N p(x_k)}, \tag{3.33}$$

since the expected number of fibers with thresholds in an interval Δx is given by the threshold distribution function as $N p(x) \Delta x$. The last term in (3.33) is of order $1/N$, so for a very large bundle, the differences among the fiber strengths in a single burst are negligible. In this sense, a burst is a *local* event.

Let us first ignore the forward and the backward conditions. Then, the number n of fibers that break as a consequence of the failure of fiber number k may take any value, and its probability distribution $\mathcal{P}(n)$ is given by a Poisson process,

$$\mathcal{P}(n) = \frac{\overline{n}^n}{n!} e^{-\overline{n}}, \tag{3.34}$$

with a parameter $\overline{n}$. Since we are interested in bursts of total size Δ, the relevant probability distribution is $\mathcal{P}(\Delta - 1)$.

The parameter $\overline{n}$ in the Poisson distribution (3.34) is the *average* number of additional broken fibers and can be determined as follows. Fiber number k carried a load $x = x_k$, and its failure causes an additional load increase

$$\delta = \frac{x}{N[1 - P(x)]} \tag{3.35}$$

on each of the remaining intact fibers. In the interval $(x, x + \delta)$ the average number of fiber thresholds is $N p(x) \delta$. The average number of fibers that break as a result of the failure of fiber number k is, therefore,

$$a(x) = N\, p(x)\, \delta = \frac{x p(x)}{1 - P(x)}. \tag{3.36}$$

If $a(x)$ were greater than unity, any failure would immediately give further failures. To avoid this proliferation of ruptures, we, therefore, assume

$$a(x) \leq 1. \tag{3.37}$$

Now $a(x) = 1$ when $1 - P(x) + xp(x) = 0$, which by (2.14) is satisfied by the critical value x_c. Thus,

$$a(x_c) = 1. \tag{3.38}$$

Therefore, the physical explanation of the condition $a(x) \le 1$ is that we are in the precritical regime where the average force $\langle F \rangle$ increases with x.

When not merely one fiber, but Δ fibers, all with approximately the same fiber strength, fail, the average number of fibers that will subsequently break is a factor Δ larger than (3.36). Thus,

$$\overline{n} = a(x)\,\Delta. \tag{3.39}$$

The probability that precisely $\Delta - 1$ additional fibers break as a consequence of the initial failure is, therefore,

$$\frac{(a(x)\Delta)^{\Delta-1}}{(\Delta-1)!}\, e^{-a(x)\Delta}. \tag{3.40}$$

We have used Eq. (3.34) with $n = \Delta - 1$ and with $\overline{n} = a(x)\Delta$, Eq. (3.39).

However, this is not sufficient. We must also ensure that both the forward condition (3.31) and the backward condition (3.32) are fulfilled.

3.3.1.1 The Forward Condition

The forward condition is a requirement to ensure that the thresholds for the additional $\Delta - 1$ fibers in the burst are not so high that the burst stops before reaching the required size Δ. Each fiber failure produces a load increase δ, given by (3.35), on the intact fibers. Thus, if the threshold x_{k+1} is less than $x_k + \delta$, fiber no. $k+1$ will be weak enough; if $x_{k+2} < x_k + 2\delta$, fiber no. $k+2$ will be weak enough; and so on. In other words, we must find at least one threshold in the interval $(x_k, x_k + \delta)$, at least 2 thresholds in $(x_k, x_k + 2\delta)$, … , at least $\Delta - 1$ thresholds in $(x_k, x_k + \Delta\delta)$.

With $x_k \equiv x$, we may state these conditions as follows. Consider a total of Δ intervals, each of size δ : $(x, x+\delta), (x+\delta, x+2\delta), \dots, (x + (\Delta-1)\delta, x + \Delta\delta)$. We must find at least one threshold in the first interval, at least two thresholds in the two first intervals, … at least $\Delta - 1$ thresholds in $(x + (\Delta-2)\delta, x + (\Delta-1)\delta)$. However, in the *last* interval $(x + (\Delta-1)\delta, x + \Delta\delta)$, we should find *no* thresholds. The reason is that we should have $F_{k+\Delta} > F_k$, Eq. (3.31), so that the fiber no. $k+\Delta$ does not rupture and thereby the avalanche is limited to size Δ. Since the burst is localized, the same *a priori* probability to find a threshold in any interval of a given size is assumed.

The combinatorial problem stated may be formulated as follows: distribute $\Delta - 1$ balls randomly in Δ numbered boxes. Find the probability p_f that there are at least h balls in the first h boxes taken together, for $h = 1, 2, \dots \Delta - 1$, while the last box is empty.

At least one ball in box number 1 implies that there should be *at most* $\Delta - 2$ balls in the last $\Delta - 1$ boxes. At most two balls in the first two boxes implies at most $\Delta - 3$ balls in the last $\Delta - 2$ boxes and so on. Thus, we seek the probability that, in general, the last p boxes contain at most $p - 1$ balls. The solution to this combinatorial problem is given in Appendix A. The resulting probability p_f that the forward condition is satisfied equals

$$p_\mathrm{f} = \frac{1}{\Delta}, \tag{3.41}$$

see Eq. (A.20).

3.3.1.2 The Backward Condition

The purpose of the backward condition is to guarantee that the burst actually starts with the failure of fiber number k. This will be the case if $F_j < F_k$ for all fibers $j < k$, Eq. (3.32). Let us begin by considering a finite number d of these fibers, namely the fibers $k-1, k-2, \ldots k-d$. If there are no thresholds in the interval $(x-\delta, x)$, at most one threshold in the interval $(x-2\delta, x)$, …, and at most $d-1$ thresholds in the interval $(x-\ \mathrm{d}\delta, x)$, then the strength requirement $F_j < F_k$ is fulfilled for all the d fibers.

In the complete interval $(x-\delta d, x)$, the probability to find h fibers is given by a Poisson distribution

$$\mathcal{P}(h) = \frac{\overline{h}^h}{h!}\, e^{-\overline{h}}, \tag{3.42}$$

with the average value $\overline{h}$ as parameter. As in (3.39), the average is given by $a = a(x)$ times the number of broken fibers, in this case

$$\overline{h} = ad. \tag{3.43}$$

Hence,

$$\mathcal{P}(h) = \frac{(ad)^h}{h!}\, e^{-ad}. \tag{3.44}$$

It remains to determine the probability that all the inequalities $F_j < F_k$ are satisfied. This is again a combinatorial problem. If the interval $(x-\ \mathrm{d}\delta, x)$ is divided into $d-1$ subintervals each of size δ, the combinatorial problem can be stated as follows: let $p(h,d)$ be the probability that by distributing h particles among d numbered boxes, so that box no. 1 contains no particles, box no. 2 contains at most 1 particle, and so on, in general that box no. i will contain at most $i-1$ particles. The solution to this combinatorial problem is given by Eq. (A.18) in Appendix A:

$$p(h,d) = 1 - h/d. \tag{3.45}$$

Summing over the possible values of h, we obtain the probability

$$p_\mathrm{d} = \sum_{h=0}^{d-1} \mathcal{P}(h)\, p(h,d) = e^{-ad} \sum_{h=0}^{d-1} \frac{(ad)^h}{h!}\left(1 - \frac{h}{d}\right) \tag{3.46}$$

that the avalanche cannot have started with any of the d fibers $k-1$, $k-2$, …, $k-d$. A slight rewriting of (3.46) gives

$$p_\mathrm{d} = (1-a)e^{-ad} \sum_{h=0}^{d-1} \frac{(ad)^h}{h!} + \frac{(ad)^d}{d!} e^{-ad}. \tag{3.47}$$

Finally, we want to take the limit $d \to \infty$. The sum in Eq. (3.47) approaches e^{ad}. In the last term in (3.47), we may use the Stirling approximation $d! \simeq d^d e^{-d}\sqrt{2\pi d}$ for large d, whereby this term takes the form $(ae^{1-a})^d/\sqrt{2\pi d}$. For $a \leq 1$, as prescribed by Eq. (3.37), this last term in (3.47) is seen to vanish in the $d \to \infty$ limit since $ae^{-a} \leq e^{-1}$. Hence, we find the probability p_b that the backward condition is satisfied:

$$p_\mathrm{b} = \lim_{d\to\infty} p_\mathrm{d} = 1 - a. \tag{3.48}$$

Summing up, we obtain the probability $\Phi(\Delta, x)$ that the fiber with threshold x is the first fiber in an avalanche of size Δ as the product of the three factors (3.40), (3.41), and (3.48),

$$\Phi(\Delta, x) = \frac{(a(x)\Delta)^{\Delta-1}}{\Delta!} e^{-a(x)\Delta}[1 - a(x)]. \tag{3.49}$$

$\Phi(\Delta, x)$ counts bursts that start with failure of a fiber with strength x. The number of fibers with thresholds in the interval $(x, x + \mathrm{d}x)$ equals $N\, p(x)\; \mathrm{d}x$. Hence, the number of bursts of size Δ and with starting point in the interval $(x, x + \mathrm{d}x)$ equals

$$\phi(\Delta, x)\; \mathrm{d}x = N\, \Phi(\Delta, x) p(x)\; \mathrm{d}x = N\, p(x)\, \mathrm{d}x\; \frac{(a(x)\Delta)^{\Delta-1}}{\Delta!} e^{-a(x)\Delta}[1 - a(x)]. \tag{3.50}$$

To find $D(\Delta)$, the total number of bursts of size Δ, irrespective of where the bursts started, we sum over all values of the starting point x,

$$D(\Delta) = \int_0^{x_\mathrm{c}} \phi(\Delta, x)\; \mathrm{d}x. \tag{3.51}$$

Thus,

$$\frac{D(\Delta)}{N} = \int_0^{x_\mathrm{c}} \frac{\Delta^{\Delta-1}}{\Delta!} a(x)^{\Delta-1} e^{-a(x)\Delta}[1 - a(x)] p(x)\; \mathrm{d}x, \tag{3.52}$$

where

$$a(x) = \frac{x p(x)}{1 - P(x)}. \tag{3.53}$$

In Sections 3.3.2 and 3.3.3, we shall study the asymptotic (large Δ) behavior of the burst distribution $D(\Delta)$, for different compositions $P(x)$ of the fiber strengths in the bundle. But first, we review some general properties of the burst distribution.

3.3.1.3 Total Number and Average Size of Bursts

What is the total number of bursts, B, in a bundle of N fibers? When a fiber fails, it will result in a failure event of *some* kind, be it an avalanche of the type we have studied or an internal burst. To avoid counting the internal bursts, we must ensure that the backward condition is fulfilled, the probability p_b of which is $1 - a(x)$ according to (3.48). The total number of bursts equals the number of failure events in which the backward condition is fulfilled:

$$B = \int_0^{x_\mathrm{c}} [1 - a(x)]\, N\, p(x)\; \mathrm{d}x = N \int_0^{x_\mathrm{c}} \left[1 - \frac{x p(x)}{1 - P(x)}\right] p(x)\; \mathrm{d}x. \tag{3.54}$$

One may obtain this result also by direct summation over the burst [22] distribution (3.52),

$$B = \sum_{\Delta=1}^{\infty} D(\Delta) = N \int_0^{x_c} \frac{1 - a(x)}{a(x)} p(x)\, s, \tag{3.55}$$

with

$$s = \sum_{\Delta=1}^{\infty} \frac{\Delta^{\Delta-1}}{\Delta!} [a(x)e^{-a(x)}]^{\Delta}. \tag{3.56}$$

In Appendix A, it is shown by the Lagrange inversion theorem that the sum s satisfies

$$s\, e^{-s} = a(x)\, e^{-a(x)}, \tag{3.57}$$

that is, $s = a(x)$. Inserting $s = a(x)$ into Eq. (3.55), the previous expression (3.54) for B is recovered.

As an example we have for the uniform distribution (2.3) of fiber strengths, a total number of avalanches equal to $B = (1 - \ln(2))N \simeq 0.304N$.

The character of these B avalanches depends very much on the loading situation. For a low load, the force fluctuations are small, and the probability to have many fibers partaking in an avalanche is small. Near the critical point, however, where the load curve for the average force is essentially horizontal, the fluctuations most effectively introduce nonmonotonicities in the F_k sequence, and consequently, large avalanches are much more probable.

We can illustrate this general behavior in two ways. A first approach is to determine the fraction q_Δ dx of all bursts that have a given length Δ and that start in the interval $(x, x + \mathrm{d}x)$. This is simply the normalized integrand of (3.52):

$$q_\Delta(x) = \frac{N}{D(\Delta)} \frac{\Delta^{\Delta-1}}{\Delta!} [1 - a(x)] a(x)^{\Delta-1}\, e^{-a(x)\Delta} p(x), \tag{3.58}$$

where $a(x)$ is given by (3.53). As shown in Figure 3.5, for the uniform fiber strength distribution, the single fiber bursts become steadily less frequent when the load increases, while bursts of greater length Δ occur more frequently at larger loads.

A second way of illustrating this general behavior is to calculate the average avalanche size $\langle \Delta \rangle$ for avalanches that start with failure of a fiber with strength x. The probability $\Phi(\Delta, x)$, Eq. (3.49), that failure of a fiber with strength x results in a burst of size Δ, depends on the burst size Δ as follows:

$$\Phi(\Delta, x) \propto \frac{\Delta^{\Delta-1}}{\Delta!} u^{\Delta}, \tag{3.59}$$

where

$$u = a(x)e^{-a(x)} = \frac{xp(x)}{1 - P(x)} \exp\left[-\frac{xp(x)}{1 - P(x)} \right]. \tag{3.60}$$

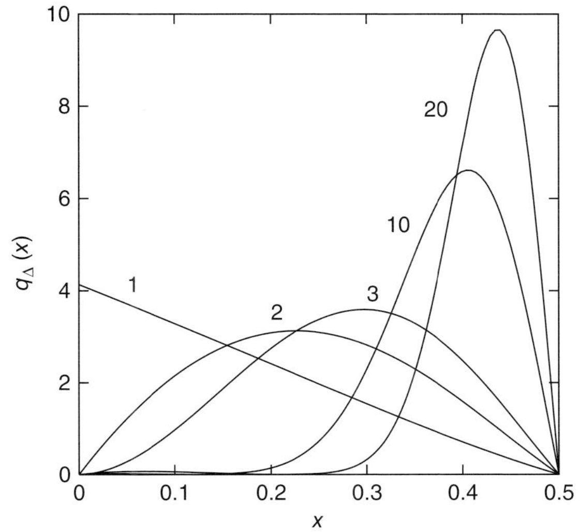

Figure 3.5 The probability density $q_\Delta(x)$ for bursts of size $\Delta = 1, 2, 3, 10$, and 20 for a fiber bundle with a uniform threshold distribution, $P(x) = x$ for $0 \le x \le 1$. The values of Δ are indicated on the graphs.

Thus, the average burst size equals

$$\langle \Delta \rangle = \frac{\sum_{\Delta=1}^{\infty} \Delta \cdot u^\Delta \Delta^{\Delta-1}/\Delta!}{\sum_{\Delta=1}^{\infty} u^\Delta \Delta^{\Delta-1}/\Delta!} = \frac{u}{s}\frac{ds}{du}, \tag{3.61}$$

where

$$s = \sum_{\Delta=1}^{\infty} \frac{\Delta^{\Delta-1}}{\Delta!} u^\Delta. \tag{3.62}$$

Inserting $u = a(x)e^{-a(x)}$ into (3.61), we obtain

$$\langle \Delta \rangle = \frac{a(x)}{1 - a(x)} s \frac{da(x)}{ds} \tag{3.63}$$

The sum s (with $u = a(x)e^{-a(x)}$) appeared in (3.56), with the result $s = a(x)$. By inserting $s = a(x)$ into (3.63), we find the average burst length equal to

$$\langle \Delta \rangle = \frac{1}{1 - s} = \frac{1}{1 - a(x)} = \frac{1 - P(x)}{1 - P(x) - xp(x)}. \tag{3.64}$$

It follows that $\langle \Delta \rangle \to 1$ for $x \to 0$, as expected. Moreover, since the critical point x_c satisfies $a(x_c) = 1$, formula (3.38), the average burst length *diverges* near the critical point. Figure 3.6 shows the variation of the average size of bursts with starting point x, for the uniform distribution and for the Weibull distribution of index $k = 3$.

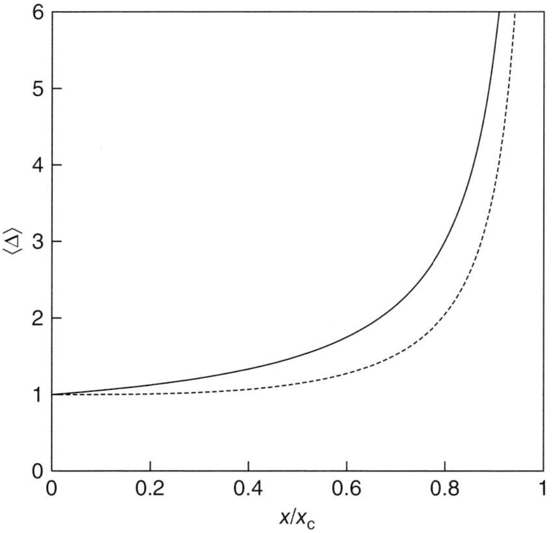

Figure 3.6 The average burst length $\langle \Delta \rangle$ as a function of x/x_c. The fully drawn graph is for the uniform distribution, $P(x) = x$ for $0 \leq x \leq 1$, for which $x_c = 1/2$. The dashed graph is for the Weibull distribution of index 3, $P(x) = 1 - e^{-x^3}$, for which $x_c = 3^{-1/3}$.

3.3.2
Asymptotic Burst Distribution: 5/2 Law

$$\frac{D(\Delta)}{N} = \frac{\Delta^{\Delta-1}}{\Delta!} \int_0^{x_c} [a(x)e^{-a(x)}]^\Delta a(x)^{-1}[1 - a(x)]\, p(x)\, dx, \tag{3.65}$$

is strongly peaked at the upper integration limit x_c since $a(x)e^{-a(x)}$ is maximal for $a(x) = 1$. Since $a(x_c) = 1$, formula (3.38), the maximum occurs at $x = x_c$. A suitable expansion of the integrand around the critical point x_c will, therefore, provide the dominating contribution for large Δ.

We expand the Δ-dependent factor in the integrand around $a = 1$ as follows:

$$a\, e^{-a} = \exp\,(-a + \ln\,[1 - (1 - a)]) = \exp\,\left[-1 - \frac{1}{2}(1 - a)^2 + \mathcal{O}(1 - a)^3\right], \tag{3.66}$$

followed by an expansion of $a(x)$ around x_c. To linear order (3.36) gives

$$a(x) \simeq 1 + a'(x_c)(x - x_c) \quad \text{with} \quad a'(x_c) = \frac{2p(x_c) + x_c p'(x_c)}{x_c p(x_c)}. \tag{3.67}$$

Insertion of these expansions into (3.65), and use of the Stirling approximation $\Delta! = \Delta^\Delta e^{-\Delta}\sqrt{2\pi\Delta}$ for the factorial, gives

$$\begin{aligned}\frac{D(\Delta)}{N} &\simeq (2\pi)^{-1/2}\Delta^{-3/2}p(x_c)\, a'(x_c) \int_0^{x_c} e^{-a'(x_c)^2(x-x_c)^2\Delta/2}\, (x_c - x)\, dx \\ &= (2\pi)^{-1/2}\Delta^{-5/2}p(x_c)a'(x_c)^{-1}[e^{-a'(x_c)^2(x-x_c)^2\Delta/2}]_0^{x_c}.\end{aligned} \tag{3.68}$$

The lower limit gives a vanishing contribution for large Δ. Hence, we obtain the following asymptotic behavior of the avalanche size distribution:

$$\frac{D(\Delta)}{N} \simeq C\,\Delta^{-5/2}, \tag{3.69}$$

where

$$C = (2\pi)^{-1/2} p(x_c)\, a'(x_c)^{-1} = \frac{x_c\, p(x_c)^2}{\sqrt{2\pi}[2p(x_c) + x_c\, p'(x_c)]}, \tag{3.70}$$

by means of (3.67).

Thus, we have obtained an asymptotic power law, $D(\Delta) \propto \Delta^{-5/2}$ for the avalanche size distribution, independent of the fiber strength mixture in the bundle. The result is *universal* for all threshold distribution for which the macroscopic force $\langle F(x)\rangle$ has a single parabolic maximum. This is the generic situation, and the restriction is necessary to ensure the validity of our expansion in $x_c - x$ near the critical point. In the next section, we study some exceptional cases where this restrictive assumption does not hold.

Note that if the experiment had been stopped before complete breakdown, at a subcritical value x_m instead of at the critical point x_c, the maximum of the Δ-dependent factor in (3.65) would be $[a(x_m)e^{-a(x_m)}]^{\Delta}$ instead of 1. In this case, the asymptotic behavior would have been dominated by an *exponential* fall-off rather than by a pure power law:

$$D(\Delta) \propto \Delta^{-5/2} e^{-[a(x_m)-1-\ln(a(x_m))]\,\Delta}. \tag{3.71}$$

Figure 3.7 shows simulation results for the avalanche distributions arising from different strain intervals $(0, x_m)$. Note how the curved distributions straighten out toward a straight line in this log-log plot when $x_m \to x_c$.

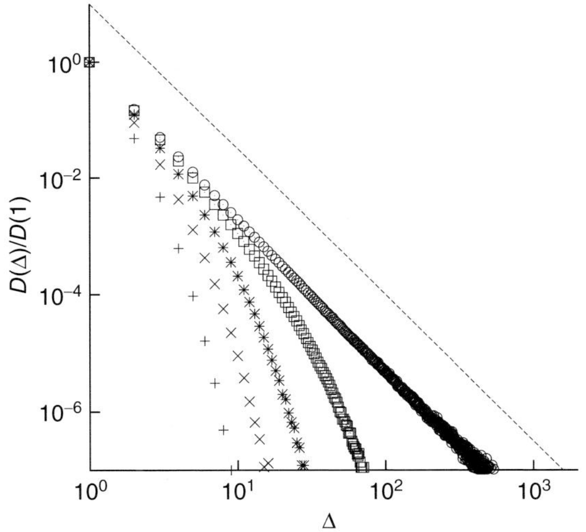

Figure 3.7 Simulation results for the normalized avalanche distribution $D(\Delta)/D(1)$, for strains in the window $(0, x_m)$. Plusses correspond to $x_m = 0.1$, crosses correspond to $x_m = 0.2$, stars correspond to $x_m = 0.3$, squares correspond to $x_m = 0.4$, and circles correspond to $x_m = 0.5$.

When x_m is close to its critical value x_c, $a(x_m)$ is close to unity, and the exponent is proportional to $(x - x_c)^2 \Delta$. The avalanche size distribution then takes the scaling form

$$D(\Delta) \propto \Delta^{-\eta}\ G(\Delta^{\nu}(x - x_c)), \tag{3.72}$$

with a Gaussian function G, a power-law index $\eta = 5/2$ and $\nu = 1/2$. This shows that the breakdown process is similar to a critical phenomenon with the critical point at total breakdown. In Chapter 5, we will meet critical exponents in a different connection.

3.3.2.1 Asymptotic Burst Distribution: Nongeneric Cases

As we have seen in Section 2.3, there may exist fiber strength distributions for which the macroscopic force $\langle F(x)\rangle$ does not have a unique parabolic maximum [18]. If there are *several parabolic maxima*, and the absolute maximum comes first (i.e., for the lowest load), the generic situation apply. If the absolute maximum does not come first, there will be several avalanches of macroscopic size. The asymptotic power law $D(\Delta) \propto \Delta^{-5/2}$ is, however, unaffected.

More interesting are threshold distributions for which the macroscopic force $\langle F(x)\rangle$ possesses no parabolic maximum at all. We are particularly interested in the resulting asymptotics of the burst size distribution.

Model examples of such threshold distributions are

$$P(x) = \begin{cases} 0 & \text{for } x \le x_0 \\ 1 - [1 + (x - x_0)/x_r]^{-\alpha} & \text{for } x > x_0, \end{cases} \tag{3.73}$$

for different sets of the positive parameters α and x_0. Here x_r is a reference quantity that, for simplicity, is set equal to unity in the following. (A special case of this threshold distribution was considered in Section 2.3, Eq. (2.24).) These distributions are all characterized by diverging moments. When $\alpha \le 1$, even the first moment – the mean – diverges. This class of threshold distributions is rich enough to exhibit several qualitatively different asymptotic size distributions for the avalanches.

The corresponding average bundle strength per fiber is, according to (2.13),

$$\langle \sigma(x)\rangle = \begin{cases} x & \text{for } x \le x_0 \\ x\,(1 + x - x_0)^{-\alpha} & \text{for } x > x_0. \end{cases} \tag{3.74}$$

Since

$$\frac{d}{dx}\langle \sigma(x)\rangle \bigg|_{x=x_0+} = 1 - \alpha x_0 < 0 \quad \text{for} \quad \alpha > 1/x_0, \tag{3.75}$$

for $\alpha > 1/x_0$ there will be an unstable part of $\langle \sigma(x)\rangle$ starting at $x = x_0$, so that $\langle \sigma(x)\rangle$ is replaced by the constant $\langle \sigma(x_0)\rangle = x_0$ in an interval from $x = x_0$ to a value $x = x_1$. In Figure 3.8, this is seen for the case $\alpha = 2/3$. The extension $x_1 - x_0$ of the flat constant-force part increases when α increases, and approaches infinity when $\alpha \to 1$.

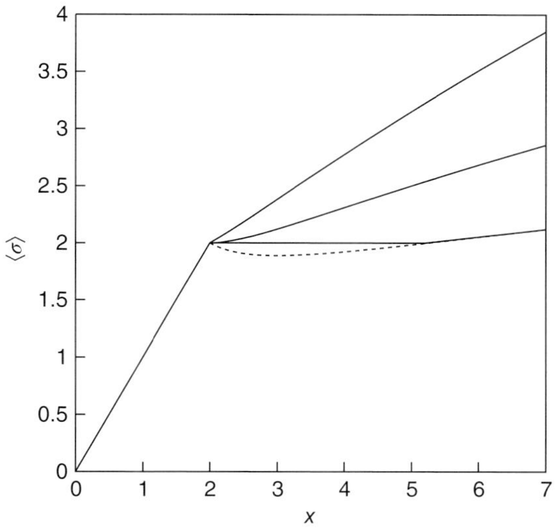

Figure 3.8 Macroscopic force curves $\langle\sigma(x)\rangle$ are sketched for the value $x_0 = 2x_r = 2$. The values of α are 1/3 (upper curve), 1/2 (middle curve), and 2/3 (lower curve). The dashed part of the $\alpha = 2/3$ curve is unstable, and the bundle strength will follow the solid line.

The distribution of avalanche sizes is given by (3.52). In the present case, the function $a(x)$, formula (3.36), takes the form

$$a(x) = \frac{xp(x)}{1 - P(x)} = \frac{\alpha x}{1 + x - x_0}. \tag{3.76}$$

A simple special case is $x_0 = 1$, corresponding to

$$p(x) = \alpha\, x^{-\alpha-1} \quad \text{for } x \geq 1, \tag{3.77}$$

then $a(x) = \alpha$, independent of x. This gives at once

$$\frac{D(\Delta)}{N} = \frac{1-\alpha}{\alpha}\,\frac{\Delta^{\Delta-1}}{\Delta!}\,(\alpha e^{-\alpha})^{\Delta} \simeq \frac{1-\alpha}{\alpha\sqrt{2\pi}}\,\Delta^{-3/2}\,(\alpha e^{1-\alpha})^{\Delta}, \tag{3.78}$$

using the Stirling approximation for the factorial. Thus, the avalanche distribution has an exponential dependence on the avalanche size Δ, modified by a $\Delta^{-3/2}$ power law.

In other cases, it is advantageous to change the integration variable in (3.65) from x to a. Using again the Stirling approximation for the factorial, we have

$$\frac{D(\Delta)}{N} = \frac{\Delta^{-3/2}}{\sqrt{2\pi}\;\alpha^{\alpha-1}(1-x_0)^{\alpha}} \int_{\alpha x_0}^{\alpha} (\alpha - a)^{\alpha-1}(a^{-1} - 1)(ae^{1-a})^{\Delta}\, da. \tag{3.79}$$

The asymptotics for large Δ, beyond the $\Delta^{-3/2}$ dependence of the prefactor, is determined by the Δ-dependent factor in the integrand. The maximum of $a\ e^{1-a}$ is unity, obtained for $a = 1$, and the asymptotics depends crucially on whether $a = 1$ falls outside the range of integration, or inside (including the border). If the maximum falls inside the range of integration, the generic $D(\Delta) \propto \Delta^{-5/2}$ dependence remains. A special case of this is $\alpha = 1$, for which the maximum of the integrand is located at the upper integration limit, corresponding to a "quadratic" maximum *at infinity*.

Another special case is $\alpha x_0 = 1$, with $\alpha < 1$, for which again the generic asymptotics $\Delta^{-5/2}$ is valid. In this case, the macroscopic force has a quadratic *minimum* at $x = x_0$ (see Figure 3.8 for $\alpha = 1/2$), and critical behavior arises just as well from a minimum as from a maximum.

In the remaining cases, in which $a = 1$ lies outside the range of integration in (3.79), the avalanche distribution is always a power law with an exponential cutoff,

$$\frac{D(\Delta)}{N} \propto \Delta^{-\xi} A^{\Delta}. \tag{3.80}$$

Here both ξ and A depend on the parameter values x_0 and α. This is easy to understand. Since

$$\frac{\mathrm{d}a(x)}{\mathrm{d}x} = \frac{\alpha(1 - x_0)}{(1 + x - x_0)^2} \tag{3.81}$$

for $x \geq x_0$, we see that $a(x)$ is a monotonically decreasing function of x for $x_0 > 1$, so that the maximum of $a\ e^{1-a}$ is obtained at the lower limit $x = x_0$, where $a = \alpha x_0$. The asymptotics

$$D(\Delta) \propto \Delta^{-5/2}\ (\alpha x_0 e^{1-\alpha x_0})^{\Delta} \tag{3.82}$$

of the avalanche size distribution follows.

This is true merely for $\alpha x_0 < 1$, however. For $\alpha x_0 > 1$, the macroscopic force $\langle f \rangle$ *decreases* near $x = x_0$, so that a macroscopic burst takes place at a force x_0 per fiber, and stabilization is obtained at a larger value x_1 (see Figure 3.7, the $\alpha = 2/3$ case). The subsequent bursts have an asymptotic size distribution of the form

$$D(\Delta) \propto \Delta^{-5/2}\ (a(x_1)e^{1-a(x_1)})^{\Delta}, \tag{3.83}$$

determined by the neighborhood of $x = x_1$.

For $x_0 < 1$, the maximum of ae^{1-a} is obtained at $x = \infty$, leading to the asymptotics

$$D(\Delta) \propto \Delta^{-\frac{3}{2}-\alpha}\ (\alpha\ e^{1-\alpha})^{\Delta}, \tag{3.84}$$

reflecting the power-law behavior of the integrand in (3.79) at infinity.

The results are summarized in Table 3.1. Note that the $x_0 = 1$ result (3.78) cannot be obtained by putting $x_0 = 1$ in (3.82), since (3.79) the order of the limits $\Delta \to \infty$ and $x_0 \to 1$ is crucial.

In conclusion, many different asymptotics may occur for the avalanche size distribution, depending on the details of the fiber strength distribution. One expects, however, that such anomalous avalanche distributions will be seldom seen. In the general case, the $D \propto \Delta^{-5/2}$ asymptotics rules.

Table 3.1 Asymptotic behavior of the avalanche size distribution for different parameter sets in the threshold distribution (3.73).

Parameters	$D(\Delta)$ asymptotics
$0 \leq x_0 < 1,\ \alpha < 1$	$\Delta^{(3/2)-\alpha}\ (\alpha e^{1-\alpha})^{\Delta}$
$0 \leq x_0 < 1,\ \alpha = 1$	$\Delta^{-5/2}$
$x_0 = 1, \alpha < 1$	$\Delta^{-3/2}\ (\alpha e^{1-\alpha})^{\Delta}$
$1 < x_0 < \alpha^{-1}$	$\Delta^{-5/2}\ (\alpha x_0 e^{1-\alpha x_0})^{\Delta}$
$1 < x_0 = \alpha^{-1}$	$\Delta^{-5/2}$
$1 < \alpha^{-1} < x_0$	$\Delta^{-5/2} e^{-\Delta/\Delta_0}$

3.3.3 Inclusive Bursts

Some bursts may occur within a larger burst, as shown in Figure 3.9. We denote bursts in which such internal bursts are included as *inclusive bursts.*

In Section 3.3.1, we avoided to include internal bursts by enforcing the *backward condition* that all forces before the start of a burst were *smaller* than the force initiating the burst. If we now want to include these internal bursts, we just delete in expression (3.52) for the burst distribution the backward condition factor (3.48),

$$p_b = 1 - a(x), \quad \text{where} \quad a(x) = \frac{xp(x)}{1 - P(x)}. \tag{3.85}$$

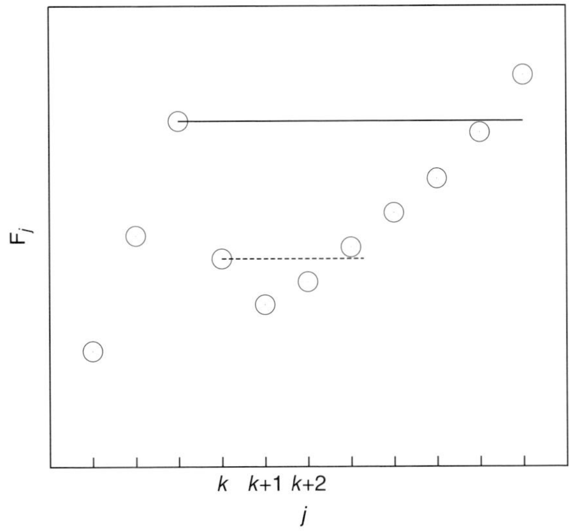

Figure 3.9 The figure shows an example of a sequence F_j of forces in which both a large burst of size 8 and an smaller internal burst of size 3 are produced.

Consequently, the distribution of inclusive bursts with burst length Δ takes the form

$$\frac{D_i}{N} = \frac{\Delta^{\Delta-1}}{\Delta!} \int_0^{x_c} a(x)^{\Delta-1}\, e^{-a(x)\Delta}\, p(x)\ \mathrm{d}x. \tag{3.86}$$

To find for large Δ the asymptotic distribution, we proceed as in Section 3.3.2: the integrand is expanded around its maximum at x_c, whereupon integration and use of the Stirling approximation give

$$\frac{D_i}{N} \simeq C_i \Delta^{-2}. \tag{3.87}$$

Here the constant $C_i = \sqrt{2\pi} C$, with C given by (3.70).

In conclusion, the inclusive burst distribution is also asymptotically a power law, with a negative exponent 2. It is not surprising that the exponent is smaller than the previous exponent 5/2, since there are so many possibilities for internal bursts within one single real burst.

The asymptotic distribution (3.87) for inclusive bursts is *universal* in the same sense as the 5/2 law for real bursts: it is the same for all threshold distributions that produce an average force $F(x)$ with a parabolic maximum.

In Figure 3.10, the inclusive distribution (3.87) for a uniform fiber strength distribution is shown to be in excellent agreement with simulation results.

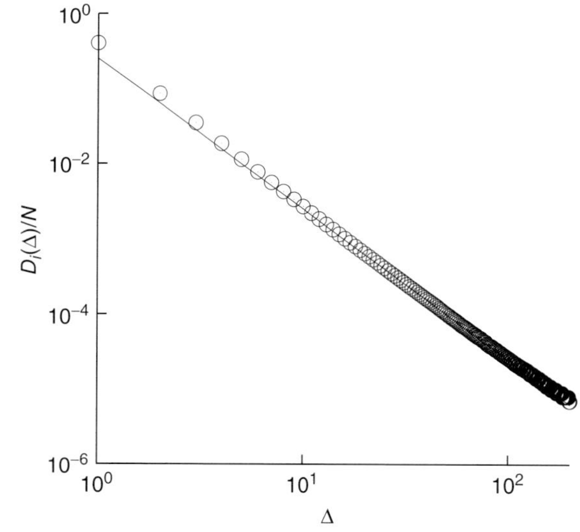

Figure 3.10 The distribution of inclusive bursts, $D_i(\Delta)/N$ for the uniform threshold distribution within 0 and 1. The straight line is a plot of Eq. (3.87). The simulation results are based on 1000 bundles with $N = 1\,000\,000$ fibers each.

3.3.4
Forward Bursts

A third kind of bursts are *forward bursts* , defined as a burst that starts at a local maximum of the fluctuating quantity and ends when a value exceeding the start value appears. Figure 3.11 shows an example.

Forward bursts have been considered in different physical contexts. An example is the pressure fluctuations observed when a fluid displaces another fluid in a porous medium [23].

The forward bursts form a subclass of the inclusive bursts, namely the inclusive bursts that start at a local maximum of the fluctuating quantity. For forward bursts, we, therefore, have to insure that the last step before the burst starts is positive, $\Delta F = F_j - F_{j-1} > 0$. In the random walk representation of the force fluctuations we found, Eq. (3.98), at extension x, the following probability distribution for the step length ΔF

$$\rho(\Delta F) = \frac{p(x)}{1 - P(x)} \exp\left[-(\Delta F + x)\frac{p(x)}{1 - P(x)}\right] \quad \text{for } \Delta F \geq -x, \tag{3.88}$$

and zero otherwise. Therefore, the probability that the step ΔF is positive equals

$$p_+(x) = \text{Prob}(\Delta F > 0) = \int_0^\infty \rho(\Delta F)\, \mathrm{d}\Delta F = \exp\left(-\frac{xp(x)}{1 - P(x)}\right) = e^{-a(x)}, \tag{3.89}$$

with $a(x)$ given by (3.85).

The burst distribution for forward bursts, $D_\mathrm{f}(\Delta)$, is given by the same distribution (3.86) as for inclusive bursts with, however, an extra factor $p_+(x)$ to ensure

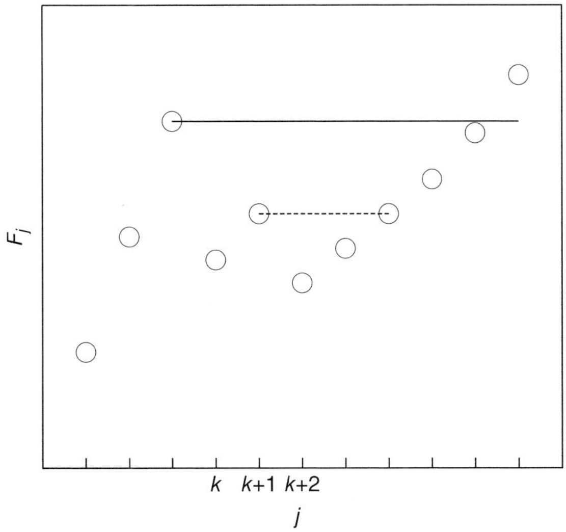

Figure 3.11 The figure shows an example of a forward burst in a sequence F_j of forces.

that the bursts start from a local maximum:

$$\frac{D_f(\Delta)}{N} = \frac{\Delta^{\Delta-1}}{\Delta!} \int_0^{x_c} a(x)^{\Delta-1} e^{-a(x)\Delta} p(x) e^{-a(x)} \, dx. \tag{3.90}$$

To determine the asymptotic distribution for large bursts, we proceed as before by expanding the integrand around its maximum at $x = x_c$. The only new feature compared with the inclusive burst distribution is the presence of the factor

$$a(x_c) = \exp\left(-\frac{x_c p(x_c)}{1 - P(x_c)} \right) = e^{-1}, \tag{3.91}$$

since $xp(x) = 1 - P(x)$ at criticality, Eq. (2.14).

Thus, the asymptotic distribution for forward bursts differs from the inclusive burst asymptotics by a factor e^{-1}:

$$D_f(\Delta)/N = e^{-1} D_i(\Delta)/N = C\sqrt{2\pi} e^{-1} \Delta^{-2}, \tag{3.92}$$

with the constant C given by (3.70). This asymptotic power law is universal.

Figure 3.12 shows the distribution of forward bursts for the uniform threshold distribution on the unit interval. The simulation results agree well with the theoretical result.

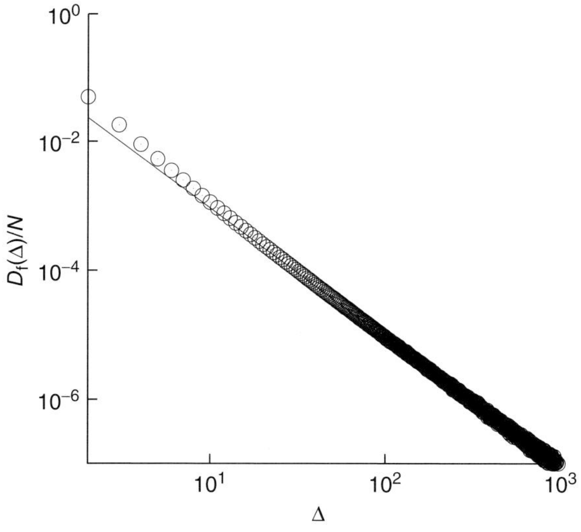

Figure 3.12 The distribution of forward bursts, $D_f(\Delta)/N$, for the uniform threshold distribution on the unit interval. The straight line shows the asymptotic distribution (3.92). The simulation results are based on 1000 bundles with N=1 000 000 fibers.

3.3.5
Avalanches as Random Walks

The ordered sequence of force values $F_k = (N + 1 - k)x_k$ when the kth fiber breaks is a stochastic process (see (3.30) and Figure 3.4) and may be considered as representing positions along an axis. The burst process can, in other words, be thought of as a one-dimensional random walk, with step lengths $F_{k+1} - F_k$ that may have either sign [24, 25].

We start by determining the exact average and variance of this unconventional random walk. Then we assume that the random walk has the same properties as an ordinary biased one-dimensional random walk with the same average, bias, and variance and use the latter to derive the burst size distribution.

3.3.5.1 The Exact Random Walk

Let us consider the probability distribution of the step length (force increase) $\Delta F = F_{k+1} - F_k$, when F_k takes place at a force $x_k \equiv x$ per fiber. Since

$$\Delta F = (N - k)(x_{k+1} - x) - x, \tag{3.93}$$

and the fiber thresholds are ordered according to strength, that is, $x_{k+1} \geq x_k$, it follows that

$$\Delta F \geq -x. \tag{3.94}$$

Let us determine the probability $q(x_{k+1})\ \mathrm{d}x_{k+1}$ to find the $(k + 1)$th threshold in the interval $(x_{k+1}, x_{k+1} + \mathrm{d}x_{k+1})$. When $x_k = x$ is given, the probability to find the values of all the remaining $N - k$ thresholds larger than x equals $[1 - P(x)]^{N-k}$. Of all these configurations, we seek those in which one of the $N - k$ thresholds has a value in the interval $\mathrm{d}x_{k+1}$ at x_{k+1}, while all the remaining $N - k - 1$ thresholds have values larger than x_{k+1}. This gives the probability

$$q(x_{k+1})\ \mathrm{d}x_{k+1} = (N - k)\frac{[1 - P(x_{k+1})]^{N-k-1}}{[1 - P(x)]^{N-k}}\ p(x_{k+1})\ \mathrm{d}x_{k+1}. \tag{3.95}$$

Using the connection (3.93), $x_{k+1} = x + (\Delta F + x)/(N + 1)$, the probability density for x_{k+1} is turned into a probability density $\rho(\Delta F)$ for ΔF:

$$\rho(\Delta F) = \frac{[1 - P(x + (\Delta F + x)/(N - k))]^{N-k-1}}{[1 - P(x)]^{N-k}}\ p\left(x + \frac{\Delta F + x}{N - k}\right), \tag{3.96}$$

properly normalized on the interval $(-x, \infty)$.

For large $N - k$, the last factor in (3.96) obviously approaches $p(x)$. In this limit, we also have

$$\begin{aligned}\left[1 - P\left(x + \frac{\Delta F + x}{N - k}\right)\right]^{N-k-1} &\simeq \left[1 - P(x) - p(x)\frac{\Delta F + x)}{N - k}\right]^{N-k-1} \\ &\simeq [1 - P(x)]^{N-k-1}\left[1 + \frac{p(x)(\Delta F + x)}{[1 - P(x)](N - k)}\right]^{N-k-1} \\ &\simeq [1 - P(x)]^{N-k-1} \exp\left[-p(x)\frac{\Delta F + x}{1 - P(x)}\right].\end{aligned} \tag{3.97}$$

Thus, the probability distribution (3.96) takes the form [26]

$$\rho(\Delta F) = \begin{cases} \frac{p(x)}{1-P(x)} \exp\left[-(\Delta F + x)\frac{p(x)}{1-P(x)}\right] & \text{for } \Delta F \geq -x \\ 0 & \text{for } \Delta F < -x. \end{cases} \tag{3.98}$$

This random walk has a very asymmetric step distribution, with steps of any size in the positive direction, while steps in the negative direction are limited in size (Figure 3.13).

The walk is in general *biased*, since the average step length

$$\langle \Delta F \rangle = \frac{1 - P(x) - xp(x)}{p(x)} \tag{3.99}$$

is nonzero, except at the critical point (see (2.14)).

The standard deviation $\delta_{\Delta F}$ of the step distribution (3.98) is easily determined via the variance

$$\delta_{\Delta F}^2 = \langle (\Delta F - \langle \Delta F \rangle)^2 \rangle = \langle (\Delta F)^2 \rangle - \langle \Delta F \rangle^2. \tag{3.100}$$

By an elementary integration, we find $\delta_{\Delta F}^2 = [1 - P(x)]^2/p(x)^2$, so that

$$\delta_{\Delta F} = \frac{1 - P(x)}{p(x)}. \tag{3.101}$$

The step bias (3.99), measured by the standard deviation, is, therefore,

$$\frac{\langle \Delta F \rangle}{\delta_{\Delta F}} = \frac{1 - P(x) - xp(x)}{1 - P(x)} = 1 - a(x), \tag{3.102}$$

with $a(x)$ defined by (3.36). Since $a(x_c) = 1$, see (3.38), the bias is zero at the critical point and otherwise positive since $a(x) \leq 1$, Eq. (3.37).

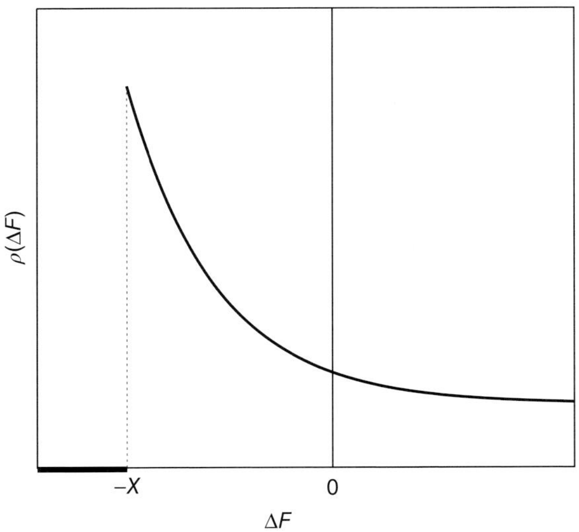

Figure 3.13 The probability distribution $\rho(\Delta F)$ of the step lengths ΔF in the exact one-dimensional random walk.

3.3.5.2 Asymptotic Burst Distribution via Random Walks

We now want to consider the burst problem in terms of the well-known one-dimensional biased random walk with a constant step length 1 [24, 25]. Let the probability to take a step in the positive direction along the z axis be p and let $q = 1 - p$ be the probability for a step in the negative direction. In each step, the average advance along the z axis is

$$\langle \Delta z \rangle_{\mathrm{rw}} = p - q, \tag{3.103}$$

and the standard deviation δ_{rw} of the step distribution is easily determined via the variance

$$\delta_{\mathrm{rw}}^2 = \langle (\Delta z)^2 \rangle - \langle \Delta z \rangle^2 = 1 - (p-q)^2 = (p+q)^2 - (p-q)^2 = 4pq. \tag{3.104}$$

Thus,

$$\delta_{\mathrm{rw}} = \sqrt{4pq}. \tag{3.105}$$

The step bias, relative to the standard deviation, is therefore

$$\frac{\langle \Delta z \rangle_{\mathrm{rw}}}{\delta_{\mathrm{rw}}} = \frac{p-q}{\sqrt{4pq}}. \tag{3.106}$$

In applying the standard random walk to the fiber burst phenomenon, we note the following:

1) As discussed in Section 3.2.1, the fiber bundle burst is a *local* event, since the fiber thresholds involved in a burst cover a tiny fraction of all fiber strength values when N is large. Therefore, the bias of the random walk can, for our purpose, be considered as *constant*.
2) We assume that the exact random walk in the fiber bundle problem can be represented by the standard random walk, provided the relative bias has the same value in the two problems,

$$\frac{\langle \Delta z \rangle_{\mathrm{rw}}}{\delta_{\mathrm{rw}}} = \frac{\langle \Delta F \rangle}{\delta_{\Delta F}} \quad \text{i.e.} \quad \frac{p-q}{\sqrt{4pq}} = 1 - a(x) = \frac{1 - P(x) - xp(x)}{1 - P(x)}. \tag{3.107}$$

3) Holding the asymptotic power law to be the most interesting feature of the burst size distribution, we consider in the present context merely the large-size asymptotics. The tail of the size distribution occurs near criticality where the bias is small. For our purpose, we may, therefore, assume the bias to be constant and *small*.

In the random walk representation, a burst of size Δ starting at position z_k may be defined as follows:

(i) $z_{k+i} \geq z_k$ for $1 \leq i \leq \Delta - 1$, and $z_{k+\Delta} < z_k$.
(ii) Furthermore, to avoid counting internal bursts (bursts within a larger burst), we must have $z_j > z_k$ for all $j < k$.

These requirements are analogs of the forward and backward conditions in Section 3.3.1. Note that the positions along the z axis have the opposite sign of the forces F_k.

The first condition is a special case of the "gambler's ruin" problem [27]: a gambler plays a series of independent games against a bank with infinite resources. In each game, the probability that the bank wins one coin is p and the probability that the gambler wins one coin is $q = 1 - p$, with $p > q$. With an initial capital of z coins, the probability $p_{\text{ruin}}(z, \Delta)$ that the gambler is ruined after precisely Δ games is straightforwardly formulated as a biased random walk problem along the z axis. For our purpose, we assume the starting point $z = 1$, and ruin corresponds to $z = 0$.

In Appendix A, it is shown that for large Δ

$$p_{\text{ruin}}(1, \Delta) \simeq \sqrt{2/\pi}\ \Delta^{-3/2}\ (4pq)^{\Delta/2}\ \sqrt{p/q}, \tag{3.108}$$

see Eq. (A.39).

Let now $p = (1 + b)/2$ and $q = (1 - b)/2$ with a small positive bias b. The connection (3.107) gives

$$1 - a(x) = b/(1 - b^2) \simeq b. \tag{3.109}$$

Thus, near criticality, the bias takes the form

$$b \simeq 1 - a(x) \simeq (x_c - x)a'(x_c), \tag{3.110}$$

where $a'(x_c)$ is given by (3.66).

For small bias, we have

$$(4pq)^{\Delta/2} = e^{\ln\ (4pq)\Delta/2} = e^{\ln\ (1-b^2)\Delta/2} \simeq e^{-b^2\Delta/2}, \tag{3.111}$$

and $(q/p)^{z/2} \simeq 1$ to lowest order.

Condition (ii) requires $z_j > z_k$ for all $j < k$, that is, that the walker arrives at z_k without ever having visited the site before. The time-reversed walk of this would be that the walker starts at z_k and never returns. The probability of no return is $|p - q|$ (see Appendix A, Eq. (A.26)).

Consequently, the probability for having a burst of a large size Δ equals

$$\Phi_{\text{rw}}(\Delta) = \frac{1}{2}|p - q|\ p_{\text{ruin}}(z = 1, \Delta) = \frac{\Delta^{-3/2}}{\sqrt{2\pi}} b\ e^{-b^2\Delta/2}. \tag{3.112}$$

Insertion of (3.110) into (3.112) gives the probability for a burst of large size Δ,

$$\Phi_{\text{rw}} = \frac{\Delta^{-3/2}}{\sqrt{2\pi}} a'(x_c)(x_c - x)\ e^{-a'(x_c)^2(x_c-x)^2\Delta/2}. \tag{3.113}$$

For the fiber bundle model, this yields the total number of bursts in the range (x_s, x_c) close to the critical point

$$\begin{aligned}
&(2\pi)^{-1/2}\ \Delta^{-3/2} p(x_c)a'(x_c) \int_{x_s}^{x_c} (x_c - x)e^{-a'(x_c)^2(x_c-x)^2\Delta/2}\ dx \\
&= (2\pi)^{-1/2}\Delta^{-5/2} p(x_c)a'(x_c)^{-1} \left[e^{-a'(x_c)^2(x-x_c)^2\Delta/2}\right]_{x_s}^{x_c} \\
&\simeq (2\pi)^{-1/2}\Delta^{-5/2} p(x_c)a'(x_c)^{-1},
\end{aligned} \tag{3.114}$$

since the contribution from the lower limit is negligible for large Δ. The result obtained is identical with the expression (3.68) for $D(\Delta)$, except that we now have restricted the integration range to the neighborhood of the critical point. However, the same asymptotic power law for the burst distribution, $\Delta^{-5/2}$, is obtained.

3.3.6 Energy Release

So far, an avalanche has been characterized by its size Δ, that is, the number of ruptured fibers in the avalanche. It may also be characterized by the amount of mechanical energy E released in the burst.

A useful experimental technique to monitor energy releases in fracture processes is to measure the acoustic emissions, the elastically radiated waves produced in the bursts. Experimental observations suggest that acoustic emission signals often follow power-law distributions, and it is natural to investigate if such power laws may result from the breaking dynamics in fiber bundles [28].

When fibers obey Hooke's law, the mechanical energy stored in a single fiber at elongation x equals $\frac{1}{2}\kappa x^2$. As before, we will normalize the elastic constant κ to unity. Since a burst is a local event, the differences in threshold values among the failed fibers in a single burst are negligible (see the discussion around (3.33)). Hence, the energy released in a burst of size Δ that starts with a fiber with threshold x is with sufficient accuracy given as

$$E = \frac{1}{2}\, x^2\, \Delta. \tag{3.115}$$

According to (3.50), the number of bursts of size Δ and starting point in the threshold interval $(x, x + \, dx)$ equals

$$\phi(\Delta, x)\, dx = N \frac{(a(x)\Delta)^{\Delta-1}}{\Delta!}\, e^{-a(x)\Delta}[1 - a(x)]\, p(x)\, dx, \tag{3.116}$$

where $a(x) = xp(x)/[1 - P(x)]$, Eq. (3.53).

The expected number of bursts with energies less than E is therefore

$$G(E) = \sum_{\Delta} \int_0^{\sqrt{2E/\Delta}} \phi(\Delta, x)\, dx, \tag{3.117}$$

with the corresponding energy density

$$g(E) = \frac{dG}{dE} = \sum_{\Delta} (2E\Delta)^{-1/2}\, \phi(\Delta, \sqrt{2E/\Delta}). \tag{3.118}$$

Explicitly,

$$g(E) = N \sum_{\Delta} g_{\Delta}(E), \tag{3.119}$$

with

$$g_{\Delta}(E) = \frac{\Delta^{\Delta-1}}{2\Delta!\, E}\, [1 - P(s) - sp(s)]\left[a(s)e^{-a(s)}\right]^{\Delta}, \tag{3.120}$$

where

$$s = \sqrt{2E/\Delta}, \tag{3.121}$$

and with the function $a(s) = sp(s)/[1 - P(s)]$ as defined by (3.53).

With a critical threshold x_c, it follows from (3.115) that a burst energy E can only be obtained for sufficiently large burst sizes Δ,

$$\Delta \geq 2E/x_c^2. \tag{3.122}$$

Hence, the sum over burst sizes in (3.119) starts with $\Delta = 1 + [2E/x_c^2]$, where $[a]$ denotes the integer part of a.

3.3.6.1 High-Energy Asymptotics

Bursts with a high energy E correspond to bursts in which many fibers rupture. In this range, we may in Eq. (3.120) use the Sterling approximation for the factorial Δ, replace the lower limit $1 + [2E/x_c^2]$ by $2E/x_c^2$, and replace the summation over Δ by an integration. Thus,

$$g(E) = \frac{N}{2(2\pi)^{1/2}E} \int_{2E/x_c^2}^{\infty} e^{\Delta}\, \Delta^{-3/2}\, [1 - P(s) - sp(s)]\, [a(s)e^{-a(s)}]^{\Delta}\, \mathrm{d}\Delta, \tag{3.123}$$

where s is given by (3.121). It is convenient to change integration variable from Δ to s, with the result

$$\begin{aligned} g(E) &= \frac{NE^{-3/2}}{2\pi^{1/2}} \int_0^{x_c} [1 - P(s) - sp(s)]\left[a(s)e^{1-a(s)}\right]^{2E/s^2} \mathrm{d}s \\ &= \frac{NE^{-3/2}}{2\pi^{1/2}} \int_0^{x_c} [1 - P(s) - sp(s)]\, e^{-Eh(s)}\, \mathrm{d}s, \end{aligned} \tag{3.124}$$

with

$$h(s) = 2[-1 + a(s) - \ln\ a(s)]/s^2. \tag{3.125}$$

For high energies E, the integral (3.124) is dominated by the contribution from the integration range near the minimum of $h(s)$. At the upper integration limit, we have $h(x_c) = 0$, since $a(x_c) = 1$, relation (3.38). For $a(x)$ near unity, we use the expansion

$$\ln\ a(s) = \ln\ [1 - (1 - a(x)] = -1 + a(x) + \frac{1}{2}(1 - a(x))^2 + \mathcal{O}[(1 - a(x))^3]. \tag{3.126}$$

Expanding also $a(x)$ in $x_c - x$ to first order in $x_c - x$, $a(x) = 1 - (x_c - x)a'(x_c)$, $h(s)$, Eq. (3.125), takes to the lowest order the form

$$h(s) = (x_c - x)^2 a'(x_c)^2\, x_c^{-2}. \tag{3.127}$$

So $h(s)$ obtains its minimum at $x = x_c$.

The first factor in the integrand of (3.124) vanishes at the critical point $s = x_c$, Eq. (2.14). Expansion of this factor to first order in $x_c - x$ gives

$$1 - P(x) - xp(x) = (x_c - x)[2p(x_c) + x_c p'(x_c)]. \tag{3.128}$$

Insertion of (3.127) and (3.128) into (3.124) gives

$$\begin{aligned}\frac{g(E)}{N} &= \frac{[2p(x_c) + x_c p'(x_c)]}{2E^{3/2}\pi^{1/2}} \int_0^{x_c} (x_c - x) \exp\left[-\frac{E(x_c - x)^2 a'(x_c)^2}{x_c^2}\right] dx \\ &= E^{-5/2} \frac{[2p(x_c) + x_c p'(x_c)] x_c^2}{4\pi^{1/2} a'(x_c)^2} \left[\exp\left[-\frac{E(x_c - x)^2 a'(x_c)^2}{x_c^2}\right]\right]_0^{x_c} . \end{aligned} \quad (3.129)$$

The lower limit zero gives a vanishing contribution for large E. By use of the result (3.67) for $a'(x_c)$, the final expression for the asymptotic energy density is

$$g(E) \simeq N \frac{C}{E^{5/2}}, \quad (3.130)$$

where

$$C = \frac{x_c^4 \, p(x_c)^2}{4\sqrt{\pi}\, [2p(x_c) + x_c p'(x_c)]}. \quad (3.131)$$

In Figure 3.14, we compare the theoretical asymptotical results with simulations for (a) the uniform distribution, $P(x) = x$ for $0 \le x \le 1$, which corresponds to $x_c = 1/2$ and $C = 2^{-7}\pi^{-1/2}$, and (b) the Weibull distribution of index 2, $P(x) = 1 - e^{-x^2}$ corresponding to $x_c = 2^{-1/2}$ and $C = 2^{-5}(2\pi\, e)^{-1/2}$.

The simulation results in Figure 3.14 for the high-energy behavior of the energy distribution function $g(E)$ are seen to be in excellent agreement with the analytic result. The theoretical derivation of (3.130) guarantees that the asymptotic behavior $g(E) \propto E^{-5/2}$ is universal (for generic forms of fiber strength distributions). Note that the exponent $-5/2$ is identical to the exponent in the asymptotic distribution (3.69) of avalanche sizes.

3.3.6.2 Low-Energy Behavior

In contrast to the energy dependence of $g(E)$ at high energies, the low-energy behavior depends crucially on the distribution of fiber strengths in the bundle. In Figure 3.15, low-energy simulation results are shown for fiber bundles with three different threshold distributions. Apparently, $g(E)$ approaches zero for $E \to 0$ in the case of a Weibull distribution of index 5, while in this limit $g(E)$ diverges for the uniform threshold distribution, and $g(E)$ approaches a constant for the Weibull distribution of index 2.

The results are easily understood. Bursts of low energy predominantly correspond to single fiber bursts ($\Delta = 1,\ E = x^2$) and to weak fibers. The number of bursts with energy less than E therefore corresponds to the number of bursts with strength $x < \sqrt{2E}$, which is close to $N\, P(\sqrt{2E})$. Hence,

$$g(E) \simeq N \frac{p(\sqrt{2E})}{\sqrt{2E}} \quad \text{when } E \to 0. \quad (3.132)$$

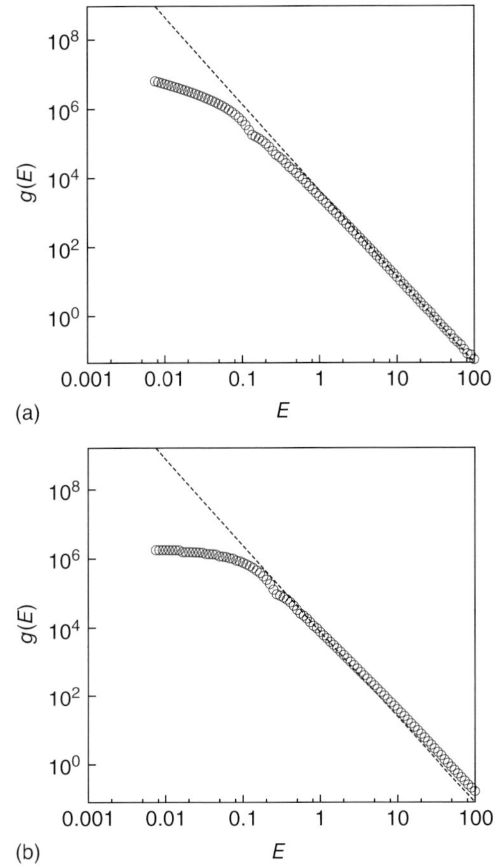

Figure 3.14 Simulation results of the energy density $g(E)$ for (a) the uniform distribution and (b) the Weibull distribution of index 2. Open circles represent simulation data, and dashed lines represent the theoretical result (3.130–3.131). In each case, the graphs are based on 1000 samples with $N = 10^6$ fibers in each bundle.

For the uniform fiber strength distribution $p(x) = 1$, and $g(E)$ therefore diverges as $(2E)^{-1/2}$ for low energies. The simulation results in Figure 3.15 are consistent with this divergence.

If the fiber strength distribution for weak fibers is linear, $p(x) \propto x$ for small x, (3.132) shows that $g(E)$ approaches a constant value for $E \to 0$. For the Weibull distribution of index k, $P(x) = 1 - \exp(-x^k)$, the density $p(x) = kx^{k-1} \exp(-x^k)$ is asymptotically linear for $k = 2$, so $k = 2$ is a borderline case between divergence and vanishing of the energy density for $E \to 0$. For the Weibull threshold distribution of index k, the low-energy behavior is $g(E) \propto E^{(k-2)/2}$, diverging for $k < 2$ and with a vanishing limit for $k > 2$, the latter result in accordance with the simulation

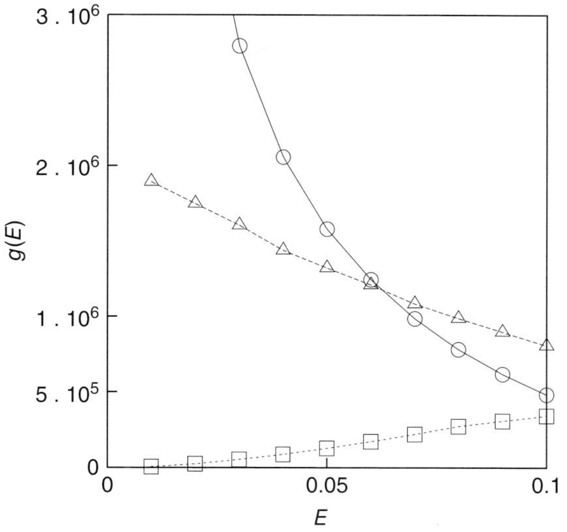

Figure 3.15 Simulation results for the energy burst distribution $g(E)$ in the low-energy regime, for the uniform threshold distribution (circles), the Weibull distribution of index 2 (triangles), and the Weibull distribution of index 5 (squares). The graphs are based on 1000 samples with $N = 10^6$ fibers in each bundle.

results in Figure 3.15. The general expression (3.120) can provide more detailed low-energy expansions.

3.3.7
Failure Avalanches for Stepwise Load Increase

So far, we have assumed that the fiber bundle is stretched continuously until complete breakdown. However, experiments may be performed in a different manner, in which the load is increased in *finite steps* of constant size δ rather than by infinitesimal amounts. The step size δ is assumed to be so large that a considerable number of fibers break in each step [29, 30].

Such a stepwise load increase is a reasonable loading method and could be a more practical loading procedure from an experimental point of view.

We record the magnitude Δ of the each burst. What may we expect for the size distribution $D_s(\Delta)$ of all avalanches up to complete bundle breakdown? Compared with the avalanche size distribution for continuous load increase, studied earlier, it is obvious that now there will be very few avalanches of small size, that is, of size smaller than δ. Avalanches of large size arose, in the previous case with continuous load increase, as a result of the force fluctuations, in particular in the critical region. The present case, however, is different. Fluctuations will be reduced, and the size of an avalanche may be large simply because each load increase will involve

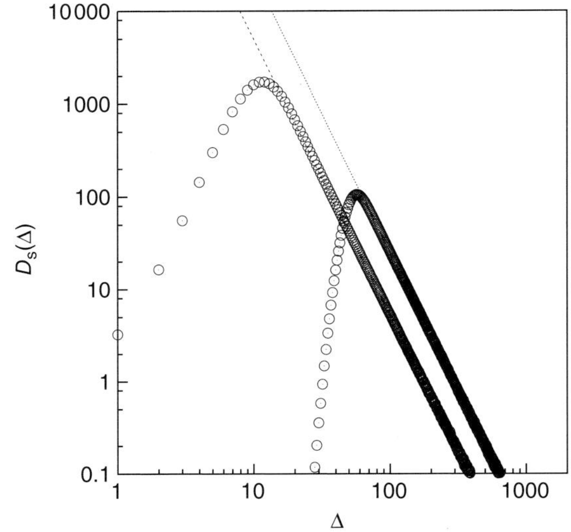

Figure 3.16 Simulation results for the avalanche size distribution $D_s(\Delta)$ when the load is increased in steps of $\delta = 10$ and $\delta = 50$ (for large avalanche sizes the graph on the right correspond to $\delta = 50$). A uniform distribution of fiber strengths is assumed. The dotted lines represent the theoretical asymptotics (3.138) for the two cases, with $D_s(\Delta) \propto \Delta^{-3}$ behavior. The graphs are based on 10 000 samples with $N = 10^6$ fibers in each bundle.

many rupturing fibers. What is the net result? In particular, how will $D_s(\Delta)$ behave asymptotically for large Δ?

Figure 3.16 shows size distributions $D_s(\Delta)$ obtained by simulation on fiber bundles with a uniform threshold distribution. The load has been increased in steps δ of size 10 and of size 50. The expected fall-off for $\Delta \leq \delta$ is present, and the large-size behavior seems to follow an asymptotic power law close to $D_s(\Delta) \propto \Delta^{-3}$ for both values of δ.

We first determine the avalanche distribution for the case shown in Figure 3.16, the uniform distribution of fiber strengths.

3.3.7.1 **Uniform Threshold Distribution**

For the uniform distribution, $p(x) = 1$ on the unit interval, zero otherwise, the load curve is parabolic,

$$\langle F(x) \rangle = Nx(1 - x), \qquad (3.133)$$

with loads between 0 and $N/4$. We therefore consider the discrete load values $m\delta$, with

$$m = 0, 1, 2, \ldots, N/4\delta. \qquad (3.134)$$

By the load curve (3.133), the threshold value x_m corresponding to the load $m\delta$ takes the value

$$x_m = \frac{1}{2}(1 - \sqrt{1 - 4m\delta /N}). \tag{3.135}$$

The expected number of fibers broken when the load is increased from $m\delta$ to $(m + 1)\delta$ is close to

$$\Delta = N \ \mathrm{d}x_m / \ \mathrm{d}m = \delta / \sqrt{1 - 4m\delta /N}. \tag{3.136}$$

The integral of (3.136) over all m from 0 to $N/4\delta$ yields a total number $N/2$ of broken fibers. This is expected, since one-half of the fibers rupture in one big final avalanche.

The number of avalanches of size between Δ and $\Delta + \mathrm{d}\Delta$ is given by the corresponding interval of the counting variable m, $D_s(\Delta)\ \mathrm{d}\Delta = \mathrm{d}m$. As (3.136) gives

$$\frac{\mathrm{d}\Delta}{\mathrm{d}m} = 2\delta^2 N^{-1}(1 - 4m\delta /N)^{-3/2} = 2\Delta^3(N\delta)^{-1}, \tag{3.137}$$

we obtain the following distribution of avalanche sizes:

$$D_s(\Delta) = \frac{\mathrm{d}m}{\mathrm{d}\Delta} = \frac{1}{2}N\delta\ \Delta^{-3}, \quad \text{for } \Delta \geq \delta. \tag{3.138}$$

For consistency, one may estimate the total number of bursts by integrating $D_s(\Delta)$ from $\Delta = \delta$ to ∞, with the result $N/4\delta$, in gratifying agreement with (3.134).

The *asymptotic* behavior of the simulation results in Figure 3.16 is in complete agreement with the theoretical formula (3.138), not only via the power law $D_s(\Delta) \propto \Delta^{-3}$, but also with respect to its amplitude.

Equation (3.136) predicts that the minimum size of the bursts is $\Delta = \delta$, obtained in the first load increase. In Figure 3.16 we see, however, ruptures of a smaller size than this. The reason is that the derivation refers to the *average* behavior, and the small-sized bursts are due to fluctuations from the average behavior.

3.3.7.2 General Threshold Distribution

Let us now consider a general threshold distribution $P(x)$, with the macroscopic load equation

$$\langle F\rangle = Nx[1 - P(x)]. \tag{3.139}$$

Again we use the discrete load values $m\delta$, with m taking integer values up to when the critical load F_c is reached. A given load $m\delta$ corresponds to a (subcritical) threshold value x_m, given by

$$Nx_m[1 - P(x_m)] = m\delta. \tag{3.140}$$

The expected number of fibers that break when the load is increased from $m\delta$ to $(m + 1)\delta$ is

$$\Delta = Np(x_m)\ \mathrm{d}x_m / \ \mathrm{d}m = \frac{p(x_m)\ \delta}{1 - P(x_m) - x_m p(x_m)}, \tag{3.141}$$

using (3.140). The number of avalanches of size between Δ and $\Delta + \mathrm{d}\Delta$ is given by the corresponding interval of the counting variable m:

$$D_s(\Delta)\ \mathrm{d}\Delta = \mathrm{d}m, \quad \text{i.e. } D_s(\Delta) = \frac{\mathrm{d}m}{\mathrm{d}\Delta}. \tag{3.142}$$

From (3.141) we obtain the derivative

$$\frac{\mathrm{d}\Delta}{\mathrm{d}m} = \delta \frac{p(x_m)[2p(x_m) + x_m p'(x_m)]}{[1 - P(x_m) - x_m p(x_m)]^2} \cdot \frac{\mathrm{d}x_m}{\mathrm{d}m} = \frac{\delta^2[2p(x_m)^2 + (1 - P(x))p'(x_m)]}{N[1 - P(x_m) - x_m p(x_m)]^3}. \tag{3.143}$$

Since $D_s(\Delta) = \mathrm{d}m/\,\mathrm{d}\Delta$, Eq. (3.142), we obtain

$$\begin{aligned} D_s(\Delta) &= \frac{N[1 - p(x_m) + x_m p'(x_m)]^3}{\delta^2\ [2p(x_m)^2 + (1 - P(x_m))p'(x_m)]} \\ &= \frac{N\delta}{\Delta^3} \cdot \frac{p(x_m)^3}{2p(x_m)^2 + (1 - P(x_m))p'(x_m)}, \end{aligned} \tag{3.144}$$

by means of (3.141).

For the uniform threshold distribution, $p(x) = 1$, this gives $D_s(\Delta) = \frac{1}{2}N\delta\Delta^{-3}$, in agreement with the previous result (3.138). In other cases x_m in (3.144) must be determined as a function of Δ through Eq. (3.141). In general, one cannot eliminate x_m explicitly, but the asymptotics can be easily found. The large-burst limit $\Delta \to \infty$ in (3.141) corresponds to $1 - P(x_m) - x_m p(x_m) \to 0$. This means that $x_m \to x_c$ for large Δ, in view of the critical condition $1 - P(x_c) - x_c p(x_c) = 0$, Eq. (2.14). Therefore, (3.144) gives the asymptotic burst-size distribution

$$D_s(\Delta) = C \cdot \Delta^{-3}, \tag{3.145}$$

where

$$C = N\delta \frac{p(x_c)^3}{2p(x)^2 + [1 - P(x_c)]p'(x_c)} = N\delta \frac{p(x_c)^2}{2p(x) + x_c p'(x_c)}. \tag{3.146}$$

In the last line we used the critical condition $1 - P(x_c) = x_c p(x_c)$.

We take the Weibull distribution of index 5, $P(x) = 1 - \exp(-x^5)$, as an example. In this case, the size distribution (3.144) and (3.141) take the following parametric form with x_m as parameter:

$$D_s(\Delta) = \frac{N\delta}{\Delta^3} \cdot \frac{25x_m^9}{4 + 5x_m^5}\, e^{-x_m^5}, \quad \text{and} \quad \Delta = \delta \frac{5x_m^4}{1 - 5x_m^5}. \tag{3.147}$$

The critical point is in this case $x_c = 5^{-1/5}$, which is the upper end point for the parameter x_m. The asymptotics is given by (3.145), with $C = N\delta(625e)^{-1/5}$.

In Figure 3.17, the size distribution function (3.147) and the asymptotic behavior are shown. These analytic results are compared with simulations using a load step $\delta = 20$. The agreement is very satisfactory.

We note that in the Weibull case, the asymptotics sets in only for rather large avalanches, compared with the results in Figure 3.16 for the uniform threshold distribution. The reason for this is obvious: on a logarithmic scale $D_s(\Delta)$ for the

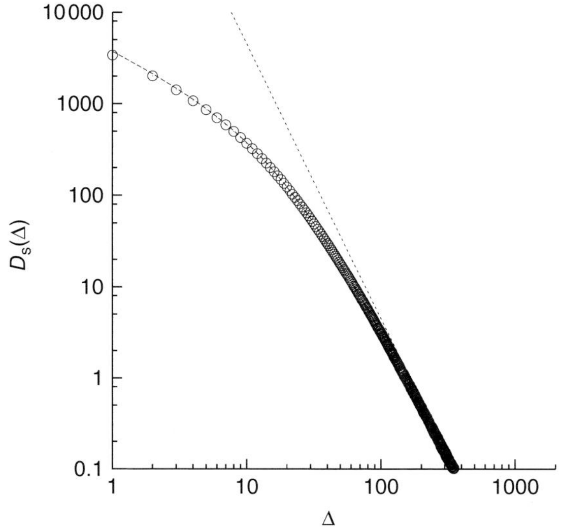

Figure 3.17 Avalanche size distribution for the Weibull threshold distribution of index 5, $P(x) = 1 - \exp(-x^5)$, with discrete load increase. The load has been increased in steps of $\delta = 20$. Open circles represent simulation data, the dashed graph is the theoretical result (3.147), while the dotted line represents the asymptotic power law with exponent −3. The simulation is based on 10 000 samples of bundles, each with $N = 10^6$ fibers.

uniform distribution, Eq. (3.138), is a linear graph all the way, and this is not so for the Weibull results. See Figures 3.16 and 3.17.

The aforementioned derivation assumes merely that the load curve has the generic parabolic form with a critical point. Granted this assumption, the asymptotic avalanche size distribution $D_s(\Delta) \propto \Delta^{-3}$ for stepwise load increase is *universal*.

4
Local and Intermediate Load Sharing

We will now enter a softer world. Our fibers have been anchored between two infinitely stiff clamps. When a fiber breaks, it distributes the load it carries to *all* the surviving fibers no matter how far they are away from it. In other words, it does not matter where the survivor is, as long as it is a survivor: it gets its share of the load inheritance from its just expired fellow fiber. Space does not enter the problem.

Is this realistic? Do fiber bundle systems in the real world behave like this? As we shall see, the answer is *yes* – in many instances they do so, but not in all. As we go along, we will refine this question and at the end of this chapter, we will not only know the answer but also have a much deeper understanding of the question.

We will consider what happens if the clamps are soft in the following. That is, they yield to the load the fibers carry. What we have in mind is perhaps nicely illustrated by pinching the hairs on your arm and pulling. The skin lifts, but not as a stiff sheet. Rather, it yields locally to the load of the individual hairs.

We show in Figure 4.1 a device that illustrates the equal-load-sharing fiber bundle model, which we have discussed until now. A number of elastic fibers are placed between two infinitely stiff clamps. By turning the handle, the two clamps are moved apart by a distance x. As the clamps are infinitely stiff, this also indicates how much each fiber is stretched.

Suppose we now exchange the infinitely stiff clamp to the right in Figure 4.1 by a clamp that has a finite stiffness – we will call it a *soft* clamp. The device now behaves as illustrated in Figure 4.2. When the clamps are moved a distance x apart, the forces carried by the individual fibers will vary depending on where the fibers are placed compared to fibers that have failed: fibers closer to holes are more extended than those farther away. Hence, they are more prone to failure under further extension than the other fibers. This effect induces a *competition* between fibers failing because they are weak – have small thresholds – and fibers that are stressed more than the other as they are close to holes.

As we shall see, the soft clamp system as shown in Figure 4.2 is tricky to deal with both analytically and numerically. We, therefore, start with a simplified model, the *local-load-sharing model.* Unfortunately, this model is also no walk in the park.

The Fiber Bundle Model: Modeling Failure in Materials, First Edition.
Alex Hansen, Per C. Hemmer and Srutarshi Pradhan.

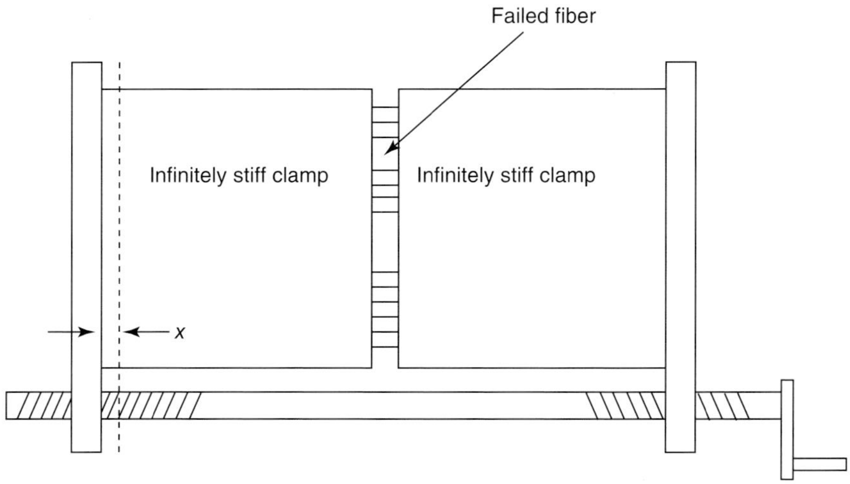

Figure 4.1 An illustration of the equal-load-sharing fiber bundle model in terms of a practical device. When the clamps are moved apart a distance x by turning the handle, the fibers will be stretched by the same amount due to the clamps being infinitely stiff.

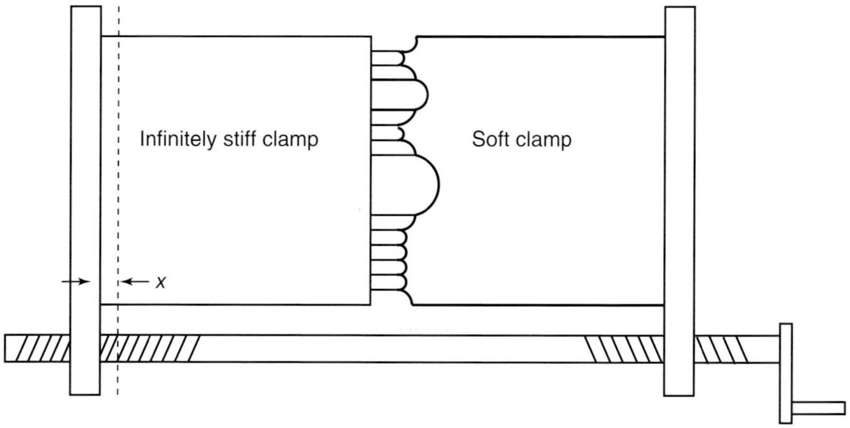

Figure 4.2 The soft clamp fiber bundle model where the fibers are placed between an infinitely stiff clamp and a soft clamp. The soft clamp responds elastically to the forces carried by the fibers. The distance between the two clamps is x as illustrated in the figure. However, the fibers are not extended accordingly as was the case in the equal-load-sharing model, see Figure 4.1.

4.1 The Local-Load-Sharing Model

The local-load-sharing fiber bundle model was first introduced by Harlow and Phoenix in 1991 [31]. The breaking rule is simple: when a fiber fails, the force it carried is distributed equally among all the *nearest* surviving fibers.

It is clear that in this model, the way the fibers are placed matters. In the following, we assume that they form a one-dimensional chain. We will later on consider two-dimensional (or more) positions. We also assume *periodic boundary conditions*, which in this one-dimensional case means that the chain forms a ring.

We illustrate the model in Figure 4.3. The soft clamp to the right in Figure 4.2 has been replaced by a clamp denoted as "hard/soft" in the local-load-sharing model. By this we mean the following: fibers that have not failed and that have intact neighbors would feel the clamp's response as if it were infinitely stiff, that is, just as in the equal-load-sharing model. When, on the other hand, fibers have failed, the clamp deforms into a *rectangular* hole as shown in Figure 4.3. This deformation is a simplification of the rounded deformation of the soft clamp in the soft clamp model. The two fibers at the edges of a given hole are stretched by an extra amount as seen in the figure. The depth of the rectangular deformation is given by the ad hoc force redistribution rule stating that "when a fiber fails, the force it carried is distributed equally among the two nearest surviving fibers" and not by any force balancing between the clamp and the fibers. Hence, the extra elongation of the fibers at the edge of a given hole is set so that the two fibers precisely carry the force that the missing fibers would have carried if they were intact.

Hence, whereas in the soft clamp model, the deformation of the clamp is due to a balancing of forces between those carried by the fibers and those transmitted through the soft clamp, in the local-load-sharing model, the deformation is solely determined by a rule specifying the force the edge fibers are to carry. However, we will later on (Section 4.3) consider the *soft membrane model*. This model follows the local-load-sharing rule as specified here when implemented in one dimension,

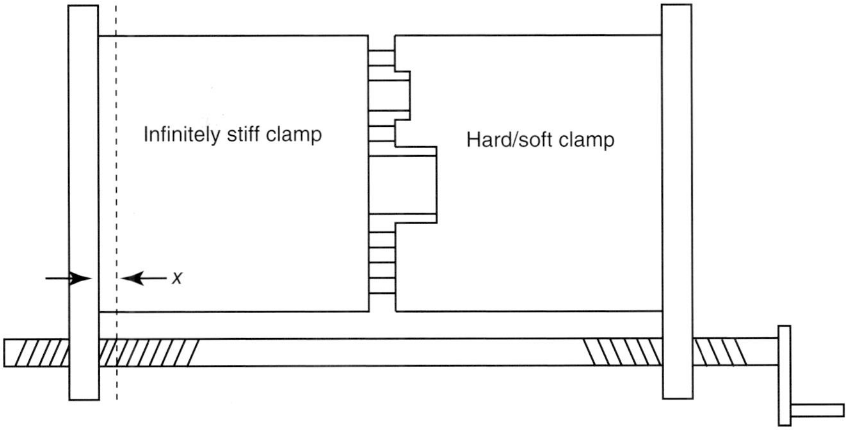

Figure 4.3 The local-load-sharing fiber bundle model illustrated with the same device as in Figure 4.2 for the soft clamp model. The soft clamp to the right in Figure 4.2 has been substituted for a clamp that reacts as an infinitely stiff clamp for the fibers that have intact neighbors. Where there are missing fibers, the clamp deforms in such a way that the fibers next to the missing ones are stretched further so that the force carried by these equals the force that would have been carried by the missing fibers. We denote this clamp as being "hard/soft."

even though the responsive clamp in this case is an elastic material (a stretchable membrane) and not a somewhat contrived construction.

4.1.1 Redistribution of Forces

Harlow and Phoenix define the local-load-sharing model as a one-dimensional array, periodic or not, of fibers. Each fiber i has a breaking threshold x_i associated with it. The thresholds are spatially uncorrelated.

We have stated the force redistribution rule as follows: when a fiber fails, the force it carried is distributed equally among the two nearest surviving fibers. Harlow and Phoenix write "In other words, the load of an initially missing or failed element is redistributed in equal portions onto its two nearest surviving neighbors."

In Kloster *et al.* [18], the model is explained as follows: "All the extra stresses are taken up by the *nearest-neighbor surviving fibers*. Precisely two fibers, one on each side, take up and divide the extra stress."

Let us analyze what this implies. Suppose we have an array of fibers all with equal spring constants $\kappa = 1$ and all carrying unit load:

$$1 \quad 1 \quad 1 \quad 1 \quad 1 \quad 1 \tag{4.1}$$

A first fiber fails, resulting in the loads

$$1 \quad 1 \quad 3/2 \quad 0 \quad 3/2 \quad 1 \tag{4.2}$$

If the left neighbor to the just-failed fiber now fails, the two intact fibers closest to the last failed fiber absorb the load that it carried. We obtain the new loads

$$1 \quad 7/4 \quad 0 \quad 0 \quad 9/4 \quad 1 \tag{4.3}$$

We see immediately that implementing the force redistribution rule as just described leads to *history dependency* of the force distribution. Had the fibers broken in the opposite order, it would have been the boundary fiber on the left that would have carried a load proportional to 9/4 and the one on the right that would have carried a load proportional to 7/4.

Hence, it is *not* possible to infer the force distribution in the model merely based on which fibers have failed. The order in which they failed has to be taken into account. This is unphysical. In spite of this, many have implemented the local-load-sharing model precisely in this way. We must, however, insist that it should be possible to determine the forces in the model from the positions of the intact fibers only.

Hence, we do not just add half the force carried by the failed fiber to each of the two nearest surviving fibers. When a fiber fails, it creates or expands a hole of failed fibers. The forces that the failed fibers would have carried is then given equally to the two surviving edge fibers.

In the aforementioned example, the force distribution would proceed as follows:

$$1 \quad 1 \quad 3/2 \quad 0 \quad 3/2 \quad 1 \tag{4.4}$$

and then

$$1 \quad 2 \quad 0 \quad 0 \quad 2 \quad 1 \tag{4.5}$$

The load carried by the fibers at the edge of the hole is determined solely by the size of the hole. There is no history dependency.

If a fiber i sits between n_l failed fibers to its left and n_r broken fibers to its right, it carries a force

$$\sigma_i = \kappa \left[1 + \frac{1}{2}\left(n_l + n_r\right)\right] x, \tag{4.6}$$

when the extension of the fiber bundle is x, where x being defined as in Figure 4.3.

The force redistribution rule (4.6) leads to a very simple relation between total force F, which is the sum of the forces carried by the surviving fibers, and the extension x. Since the fibers at the edges of the holes absorb the force that should have been carried by the failed fibers in the holes, we have

$$F = \kappa N x \tag{4.7}$$

throughout the entire failure process. Hence, the force per fiber (intact or not) is

$$\sigma = \frac{F}{N} = \kappa x, \tag{4.8}$$

throughout the entire failure process. We repeat ourselves in emphasizing once again that the definition of x is given in Figure 4.3.

4.1.2
Determining the Failure Sequence

In the earlier chapters, which focused on the equal-load-sharing model, the simplicity of the rule by which the fibers fail (which is "the weakest remaining fiber fails") made it unnecessary to discuss this rule any further. This situation has now changed. In the local-load-sharing model, as in other soft clamp models, fibers may fail as a result of being weak but also as a result of being highly stressed. Hence, we need to find a rule by which the fibers fail that takes this into account.

Each fiber i has a failure threshold x_i, and it carries a force $\sigma_i(x)$ given by Eq. (4.6) for an extension x.

Suppose we start at $x = 0$ with all fibers intact. As x is increased, a first fiber – the one with the smallest threshold – fails. Instantly, the force it carried is redistributed. How are we to determine what happens next? We may proceed as follows: as soon as the first fiber has failed, we reduce the extension x back to zero and then start increasing it until a second fiber fails. The procedure is then repeated with a reduction of x back to zero after each new failure. This is *quasistatic loading.*

But, this procedure as described so far does not tell us which fiber will fail next after the first fiber has failed. It is no longer necessarily the one with the smallest threshold among the remaining fibers since the forces are now unevenly distributed: two fibers now carry a load of $(3/2)\kappa x$, whereas the others carry a force κx each.

We define in the following the procedure by which we determine which fiber will fail next when the loading is quasistatic. This description may seem unnecessarily complicated at this point, but the reader should have patience: when we further on in this chapter deal with very complicated force redistribution mechanisms such as in the soft clamp model, the procedure we now are about to describe can be taken over unchanged.

Let us return to the very beginning of the process when all fibers are intact. We then look for the smallest threshold, or equivalently

$$\max_i \left(\frac{1}{x_i} \right). \tag{4.9}$$

The load on fiber i is $\sigma_i(x) = \kappa x$. As x is increased from zero, it will be the fiber with the smallest threshold that reaches the equality $\sigma_i(x) = \kappa x = \kappa x_i$ first whereas the other fibers will still fulfill the inequality $\sigma_i(x) = \kappa x < \kappa x_i$. We combine these two expressions with Eq. (4.9),

$$\max_i \left(\frac{x}{x_i} \right) = 1, \tag{4.10}$$

or

$$\max_i \left(\frac{1}{x_i} \right) = \frac{1}{x}. \tag{4.11}$$

We now note that $\sigma_i(x = 1) = \sigma_i(1) = \kappa$ and rewrite this equation as

$$\max_i \left(\frac{\sigma_i(1)}{x_i} \right) = \frac{\kappa}{x}. \tag{4.12}$$

This last expression is directly applicable for determining the subsequent failures of the fibers, also when the forces redistribute themselves unevenly. One determines the largest ratio $\sigma_i(x = 1)/(\kappa x_i)$ among the surviving fibers. This ratio is the inverse of the extension at which this fiber fails. It works because between each failure event, the model is linear and one has $\sigma_i(x) = \sigma_i(1)x$ – even if $x < 1$. It simply determines which fiber that will reach the condition $\sigma_i(x) = \kappa x_i$ first as x is increased from zero, and the extension at which this fiber fails.

The force redistribution rule (4.6) and the breaking rule (4.12) define the model completely. We are now ready to study it. We set the fibers' spring constant $\kappa = 1$ in the following.

4.1.3 Bundle Strength

As for the equal-load-sharing model, the most important quantity we want to know about the local-load-sharing model would be the critical extension of the

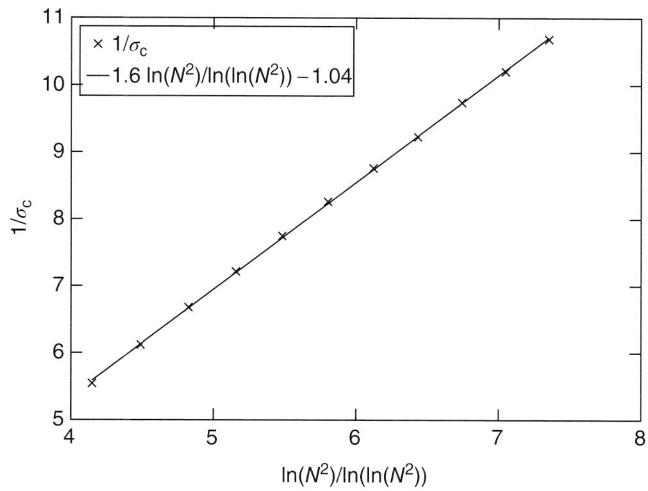

Figure 4.4 Inverse of critical load per fiber $1/\sigma_c$ versus $\ln(N^2)/\ln(\ln(N^2))$ based on 10^5 samples for $N = 100$ up to 2000 samples for $N = 102\,400$. The threshold distribution was uniform on the unit interval.

bundle, x_c, or equivalently, through Eq. (4.7) the critical load per fiber $\sigma_c = F_c/N = x_c$. One could imagine that σ_c would be rather independent of the number N of fibers in the bundle, since the average critical extension of the fibers does not change when N is increased. Such a behavior would be like the result for the equal-load-sharing model, but it is incorrect here for the one-dimensional local-load-sharing model, as we see now (but it *is* correct for the local-load-sharing model in higher dimensions, see Section 4.2.4.).

In Figure 4.4, we have simulated the critical strength for several values of N for the case of a flat distribution of fiber thresholds. The results correspond to

$$\sigma_c^{-1} \propto \frac{\ln(N^2)}{\ln(\ln(N^2))} \tag{4.13}$$

for large N. Thus, in contrast to the equal-load-sharing case, the critical force increases slower than being proportional to N.

This is in one dimension where the fibers are placed along a line. When the fibers form a two-dimensional pattern, the result is surprisingly different as we shall see in Section 4.2.4.

The cause of the behavior in one dimension must be that the probability to find somewhere a stretch of failed fibers that starts a fatal burst increases with the bundle size. To study this, we must investigate the failure mechanism in more detail, and we start with the simplest case.

4.1.4
Failure of First and Second Fibers

What is the probability p_i that a given fiber i fails first when they are all intact in a local-load-sharing model consisting of N fibers? It is simply $p_i = p = 1/N$.

As soon as a first fiber has failed, the forces on the fibers are no longer equal, and we now pose the following question: what is the probability that the second fiber to fail is one of the two neighbors of the fiber that failed first? This probability is equivalent to the probability that after two failed fibers, there is one hole consisting of two adjacent failed fibers. This probability $p_{\rm h}(2,2)$ is a special case of the *hole size distribution* $p_{\rm h}(m,k)$, which is the average fraction of failed fibers found in holes of size m when k fibers have failed. We calculate in the following $p_{\rm h}(2,2)$ as a function of N assuming that the threshold distribution is uniform on the unit interval. We will then generalize it to arbitrary threshold distributions.

One may think that $p_{\rm h}(2,2)$ is $3/N$ for large N since each of the two adjacent fibers to the one that failed first has an increased probability of breaking equal to $3/2$ due to the increased force they are carrying. As there are two of them, the probability becomes $2 \times (3/2)/N$, which is $3/N$. In Figure 4.5, we show numerical calculation of $p_{\rm h}(2,2)$ for N in the range 16-256. We find $p_{\rm h}(2,2) = 4/N$ rather than $3/N$.

The argument giving $3/N$ is simple and seems iron clad. Where does it fail? We must take into account that when removing the fiber with the smallest threshold, we have in fact changed the distribution from which the remaining $N-1$ fibers were drawn: it is no longer the uniform distribution on the unit interval. One may argue that removing one single fiber cannot create much of a change for the remaining fibers. However, the breaking criterion (4.12) is focused on the remaining *small* thresholds, which are precisely those for which the removal of the smallest threshold matters the most.

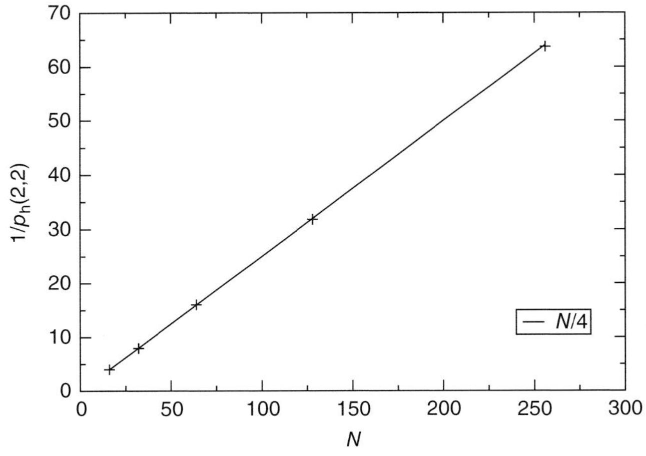

Figure 4.5 The inverse of the probability $p_{\rm h}(2,2)$ to find a hole of size 2 when two fibers have failed as a function of N for the threshold distribution $P(t) = t$ for $0 \leq t \leq 1$. The data points are based on 5×10^5 samples each. The straight line is $N/4$.

In fact, the calculation of $p_{\mathrm{h}}(2, 2)$ is an example of a problem within the realm of order statistics (see Appendix B.2).

We focus on a given fiber bundle with thresholds x_i, $i = 1, \dots, N$. We order the indexing of the thresholds so that $x_1 < x_2 < \cdots < x_N$.

Suppose that the first fiber has failed. We are then left with the sequence of thresholds $x_2 < x_3 < \cdots < x_N$. Suppose that the fiber with threshold x_k is next to the failed fiber. This fiber carries a force that is 3/2 times that of the other fibers that are not next to the failed fiber. Hence, if this fiber is to fail next, it has to

- have a threshold x_k that is smaller than $(3/2)x_2$ and
- have a threshold that is smaller than the threshold of the other fiber next to the fiber that failed first.

We consider the first of the two conditions. We need to calculate

$$\pi_k = \mathrm{Prob}\left(x_k \leq \frac{3}{2} \quad x_2\right). \tag{4.14}$$

From Appendix B.3, we have that the joint probability distribution for x_m and x_l is given by Eq. (B.29) [32].

$$\phi_N(x_m, x_l) = C P(x_m)^{m-1}(P(x_l) - P(x_m))^{l-m-1}(1 - P(x_l))^{N-l} p(x_m) p(x_l),$$

when $x_m < x_l$. The combinatorial factor C is given by Eq. (B.29),

$$C = \frac{N!}{(m-1)!(l-m-1)!(N-l)!}.$$

$P(x)$ is the cumulative threshold probability and $p(x) = \mathrm{d}P(x)/\mathrm{d}x$ is the corresponding probability density. We are interested in the case when $m = 2$ and $l = k$ as we wish to calculate the probability (4.14). Hence,

$$\phi_N(x_2, x_k) = C P(x_2)(P(x_k) - P(x_2))^{k-3}(1 - P(x_l))^{N-k} p(x_2) p(x_k), \tag{4.15}$$

with

$$C = \frac{N!}{(k-3)!(N-k)!}. \tag{4.16}$$

We calculate the probability π_k (4.14) by integrating over the gray area in Figure 4.6, that is,

$$\pi_k = \int_0^1 \mathrm{d}x_k \int_{2x_k/3}^{x_k} \mathrm{d}x_2 \quad \sigma_N(x_2, x_k). \tag{4.17}$$

When the cumulative threshold probability is $P(x) = x$ for $0 \leq x \leq 1$ – that is, the uniform distribution on the unit interval – we find

$$\pi_k = \frac{2k-1}{3^{k-1}}. \tag{4.18}$$

Surprisingly, there is no N dependency in this expression.

We now turn to the second condition, namely that threshold x_k is the smaller of the thresholds of the two fibers neighboring the fiber that failed first. With $N - 1$ surviving fibers, they may be placed in $(N - 1)!$ different configurations.

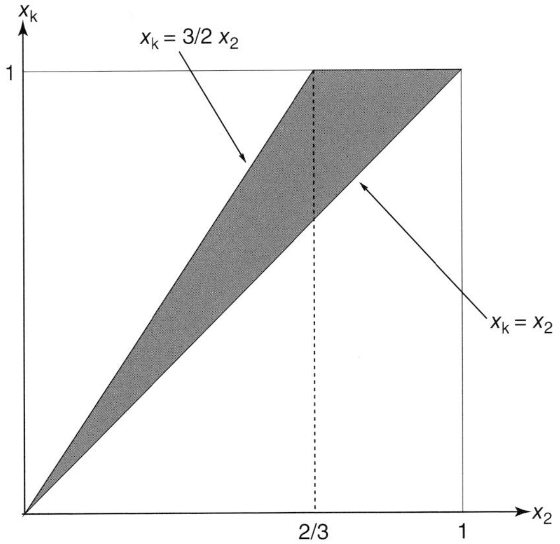

Figure 4.6 The integration area used in calculating π_k in Eq. (4.17).

In $2(N-2)!$ of these configurations, x_k will be a neighbor of the fiber that failed first. Of these $2(N-2)!$ configurations, there will be $2(k-2)(N-3)!$ configurations where it is not the smaller. We combine these factors and find

$$C_k = \frac{2(N-2)! - 2(k-2)(N-3)!}{(N-1)!} = \frac{2(N-k)}{(N-1)(N-2)}, \tag{4.19}$$

where C_k then denotes the fraction of the $(N-1)!$ configurations that x_k is the smaller of the thresholds of the two fibers neighboring the one that failed first.

We now combine Eqs. (4.18) and (4.19) to find $p_{\mathrm{h}}(2,2)$:

$$p_{\mathrm{h}}(2,2) = \sum_{k=2}^{N} C_k \pi_k. \tag{4.20}$$

We calculate the sum and find

$$p_{\mathrm{h}}(2,2) = \frac{27 + 18N - 23\quad 3^N + 8N3^N}{2(N-1)(N-2)3^N}. \tag{4.21}$$

We compare in Figure 4.7 this expression with direct numerical simulation. The agreement is perfect.

When we let $N \to \infty$ in Eq. (4.21), we find

$$p_{\mathrm{h}}(2,2) = \frac{4}{N}, \tag{4.22}$$

as expected from Figure 4.5.

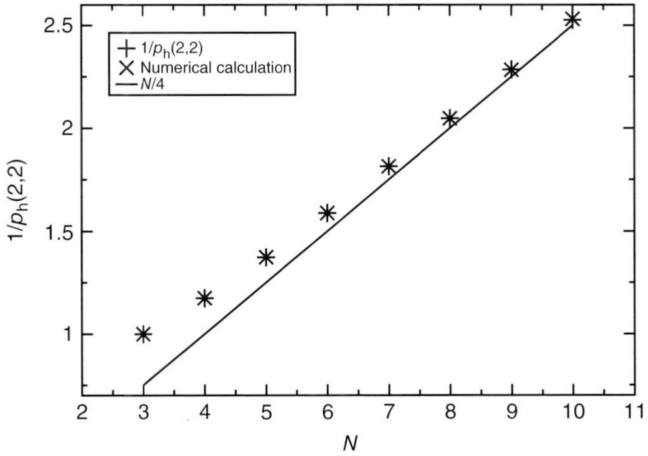

Figure 4.7 $1/p_h(2,2)$ as a function of N from Eq. (4.21) compared with numerical simulations based on 10^6 samples for each N. We furthermore compare with the asymptotic expression $N/4$.

Emil Julius Gumbel

© University Archives Heidelberg, Germany

Emil J. Gumbel was born in 1891 into an affluent family in Munich. He attended the University of Munich 1910–1914, taking courses in mathematics, economics, and social sciences. In 1914, he was awarded his doctorate in statistics, with highest honors. Eight days after he had received his diploma, World War I broke out. Gumbel volunteered military service in August 1914. After a training period, he served in the First Bavarian Ski Battalion, but was soon dismissed for health reasons, a severe respiratory infection.

The death of Gumbel's cousin in battle and the ultranationalistic attitude of most German professors stimulated him to reconsider his attitude to the conflict. He supported the peace movement and emphasized international scholarly solidarity. In 1915, Gumbel moved to Berlin. He enrolled at the University of Berlin, where he worked with von Bortkiewicz, the leading German statistician of the era. Gumbel became more and more involved in political activities. He became the leader of the Berlin section of the International Student Association and spearheaded a small campaign in support of a Munich professor in trouble because of his article concerning Germany's war guilt. The agitation irritated the University leadership, and its rector met several times with the 25-year-old Gumbel in a vain effort to silence the International Student Association. This was the first of the many encounters between Gumbel and the conservative academic authorities. The most active phase of Gumbel as an advocate of international understanding ended in the 1920s.

From then on he focused on domestic affairs. Gumbel concentrated on the perpetration of political terror in Germany by paramilitary groups and secret societies. He wrote extensively in newspapers, journals, and books: *Conspirators* (published in Vienna) and *Two years of political murder* (followed later by a second edition entitled *Four years of political murder*). These books were important in raising these issues to public consciousness. In 1924, the Ministry of Justice investigated Gumbel on the charge of treason. Gumbel had obtained a position as outside lecturer (Privaldozent) in statistics at the University of Heidelberg. However, in 1924, the Heidelberg professors, with one dissenting vote, elected to dismiss him from his post. The reason was that Gumbel, at a rally sponsored by several pacifist groups to commemorate the tenth anniversary of the outbreak of the war, uttered words that were construed as an insult to the war victims and to the German nation: "...they had died on the fields of dishonor." An investigating committee was formed, and witnesses, including war veterans, were heard. The committee asked for a broader study of Gumbel's personality to investigate Gumbel's fitness to teach. The "Gumbel affair" created a heated debate and was reported also outside Germany. A rally gathering 10 000 participants was held in Heidelberg demanding the dismissal of Gumbel. Finally, the Baden Ministry of Education did not fire Gumbel, but he remained a target for anti-pacifist, anti-socialistic, and anti-Semitic sentiments.

In 1930 Gumbel was promoted to extra-ordinary professor at Heidelberg, at the same time as the stunning electory victory of the Nationalist Party. The party was successful in winning support among students, and in November 1930 two thousend Nazi students held a public rally against Gumbel. He was dismissed in June 1932, the first professor fired because of Nazi pressure. He was also one of the 33 first Germans stripped of their citizenship by the Hitler regime. Gumbel moved to France and lectured in Lyon from 1934 to 1940. In the late 1930s he shifted his research from social questions to statistical analysis of natural and physical phenomena. He developed the theory of extreme values, and applied it to floods and to the breaking strength of materials.

Gumbel became a French citizen in 1939, but the German invasion of France made his position vulnerable. At the initiative of Albert Einstein he obtained an invitation to join the New School for Social Research in New York, and arrived in New York in October 1940. He stayed at this institution for four years. Later on he had a position at Brooklyn College, but lost his job after he failed an entire class in elementary statistics. However, Gumbel's growing scholarly repute led to several invitations and consulting contracts. In 1953 he was appointed adjunct professor at Columbia University, where he worked for the rest of his life. He made very important contributions to statistics and extreme statistics. Gumbel's name is associated with one of the three asymptotic distributions in extreme statistics (see appendix B.3.1).

The crowning work of Gumbel's career was his book from 1958, Statistics of Extremes, which immediately won high regard in the international scientific community. Gumbel died in 1966.

4.1.4.1 Other Threshold Distributions

Equation (4.21) was calculated under the assumption that the threshold distribution was uniform on the unit interval. We now consider a general threshold distribution $P(x)$ where $x_< \leq x \leq x_>$. The probability π_k, defined in Eq. (4.14), is then given by

$$\pi_k = \int_{x_<}^{x_>} \mathrm{d}x_k \int_{2x_k/3}^{x_k} \mathrm{d}x_2 \;\phi_N(x_2, x_k). \tag{4.23}$$

As an example, let us assume $P(x) = x^\beta$ for $0 \leq x \leq 1$. With $\beta = 1$, we have the uniform distribution on the unit interval, which we have studied already. The integral (4.23) splits into the product of two simple integrals if we make the substitution $x_2 = x_k x$ where $2/3 \leq x \leq 1$,

$$\pi_k = C\beta^2 \left[\int_0^1 \mathrm{d}x_k \;\; x_k^{k\beta}\left(1 - x_k^\beta\right)^{N-k}\right]\left[\int_{2/3}^1 \mathrm{d}x \;\; x^{2\beta-1}\left(1 - x^\beta\right)^{k-3}\right] \tag{4.24}$$

giving

$$\pi_k = \left(1 - \left(\frac{2}{3}\right)^\beta\right)^{k-2}\left(1 + (k-2)\left(\frac{2}{3}\right)^\beta\right), \tag{4.25}$$

which reduces to Eq. (4.18) when $\beta = 1$.

We now perform the sum in (4.20) and find

$$p_\mathrm{h}(2,2) = \frac{2^{1-2\beta}}{(N-1)(N-2)}\left[3^{2\beta}\left(1 - \left(\frac{2}{3}\right)^\beta\right)^{N-1}\left(3 + \left(\frac{2}{3}\right)^\beta (N-3)\right)\right.$$
$$\left. -3^{1+2\beta} - 2^{2\beta}(N-1) + 2^{1+\beta}3^\beta N\right]. \tag{4.26}$$

In the limit of large N, this expression becomes

$$p_\mathrm{h}(2,2) = \frac{2}{N}\left[2\left(\frac{3}{2}\right)^\beta - 1\right], \tag{4.27}$$

which reduces to $4/N$ – Eq. (4.22) – in the uniform distribution limit $\beta = 1$.

It is interesting to consider the $\beta \to 0$ limit. We see from Eq. (4.25) that in this limit $\pi_2 = 1$ and $\pi_k = 0$ for $k > 2$. Hence,

$$p_\mathrm{h}(2,2) = \frac{2}{N-1}, \tag{4.28}$$

which is what would be expected in the equal-load-sharing fiber bundle model. This is not accidental. The $\beta \to 0$ limit is an *infinite-disorder* limit. By this we mean the following. We have the N threshold values $x_1 \leq x_2 \leq \cdots \leq x_N$. We average the nth threshold in this sequence over many samples, $\langle x_n \rangle$. From order statistics (Appendix B.2), we have that (B.28)

$$\frac{\langle x_{n+1}\rangle}{\langle x_n \rangle} = \left(\frac{n+1}{n}\right)^{1/\beta}. \tag{4.29}$$

This ratio diverges as $\beta \to 0$. Hence, when this ratio diverges, it will not matter whether x_2 is enhanced by a factor 3/2 compared to the other thresholds or not.

As a result, the behavior is the same as that of the equal-load-sharing model. The same argument applies to all $p_h(m,k)$: they will all be equal to those of the equal-load-sharing model. The local-load-sharing fiber bundle model crosses over to the equal-load-sharing fiber bundle model in the infinite disorder limit.

4.1.4.2 Localization

Let us now consider what happens if there is an $x = x_< > 0$ so that $P(x) = 0$ for $x \le x_<$. A simple choice for the threshold distribution is

$$P(x) = 1 - e^{-(x-x_<)/\tau} \quad \text{for } x_< \le x < \infty, \tag{4.30}$$

We calculate π_k from Eq. (4.17) to find

$$\pi_k = 1 - I_b(N+1-k, k-2) - \frac{b^{-2N}}{B_1(N+1-k, k-2)}$$
$$(Nb^2 B_b(3N-k-1, k-2) - (N-1)B_b(3N-k+1, k-2)) , \tag{4.31}$$

where $b = \exp\,(-x_</2\tau)$, $B_b(a,c) = \int_0^b \mathrm{d}x \quad x^{a-1}(1-x)^{c-1}$ is the incomplete Euler beta function and $I_b(a,c) = B_b(a,c)/B_1(a,c)$. Using Eq. (4.20), we then construct $p_h(2,2)$ and find the surprisingly simple result

$$p_h(2,2) = 1 - \frac{N-3}{N+1} \quad e^{-x_</\tau}. \tag{4.32}$$

We show in Figure 4.8 this equation compared to numerical calculations for $N = 5$ and $N = 500$ for different $x_</\tau$. There is no appreciable difference between Eq. (4.32) and the numerical simulations.

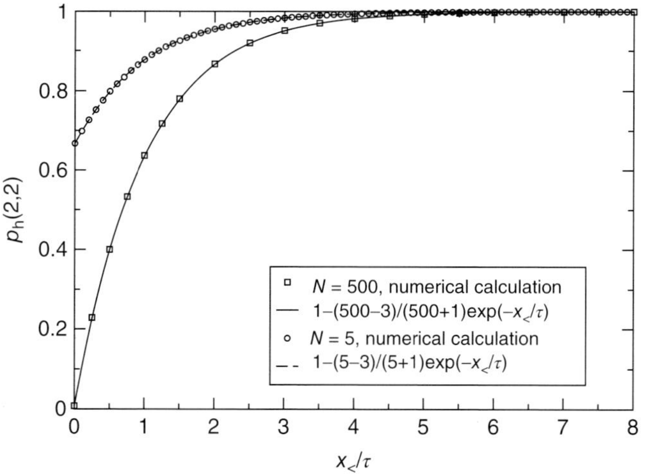

Figure 4.8 $p_h(2,2)$ for the exponential threshold distribution, Eq. (4.32) compared with numerical calculations for $N = 5$ and $N = 500$. The statistics is based on 10^6 samples for each curve.

It is interesting that Eq. (4.32) predicts

$$p_{\mathrm{h}}(2,2) = \frac{4}{N+1}, \tag{4.33}$$

when $x_< = 0$, irrespective of the value of τ. Hence, it shows the same behavior as when the threshold distribution is uniform on the unit interval, see Eq. (4.22) when $N \to \infty$. Is this surprising? No. Since $p_{\mathrm{h}}(2,2)$ is determined by the very small thresholds, it does not see the difference between the uniform distribution $p(x) = 1$ and the exponential distribution $p(x) = \exp(-x/\tau)/\tau = (1 - x/\tau + \cdots)/\tau$.

We now return to the case when $x_< > 0$. Equation (4.32) implies that $p_{\mathrm{h}}(2,2) \to 1$ as $x_< \to \infty$. The approach of $p_{\mathrm{h}}(2,2)$ to one is exponentially fast. Already for $x_</\tau = 1$, the probability to create just one hole is about 0.63. If this happens, the enhancement of force in the two fibers now at the edge of the size 2 hole is even bigger than when the hole had size 1. Hence, it is very likely that the next fiber to fail will be one of the edge fibers. If this happens, the enhancement of force carried by the edge fibers increases even further and so on. We end up with adjacent fibers failing consecutively. This phenomenon is *localization.* This is a mechanism that is absent in the equal-load-sharing fiber bundle.

Let us now consider a threshold distribution that has a finite upper cutoff $x_>$ in addition to the lower cutoff $x_<$. The simplest case to consider is a uniform distribution $1/(x_> - x_<)$ between the two cutoffs. The corresponding cumulative probability is

$$P(x) = \frac{x - x_<}{x_> - x_<} \quad \text{for } x_< \leq x \leq x_>. \tag{4.34}$$

For simplicity, but without loss of generality, (as we may always rescale the results), we set $x_> = 1$. As the expression for $p_{\mathrm{h}}(2,2)$ is rather complicated, we restrict ourselves to plot it in Figure 4.9. When $x_<$ reaches the value $2/3, p_{\mathrm{h}}(2,2) = 1$. Hence, it is guaranteed that the adjacent fibers fail consecutively, that is, the failure is localized. We may understand this result physically by the following line of

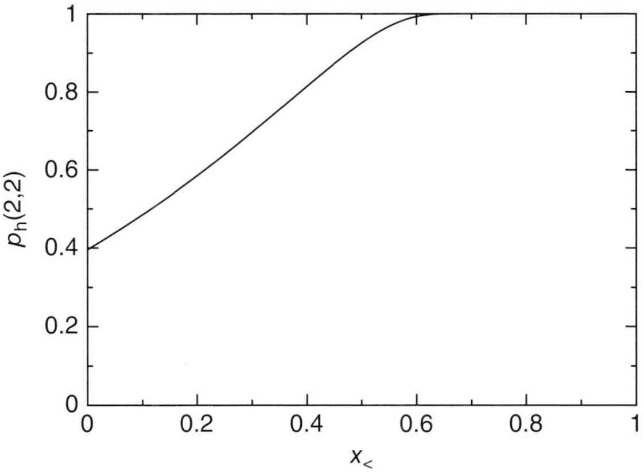

Figure 4.9 $p_{\mathrm{h}}(2,2)$ for the threshold probability given in Eq. (4.34) as a function of the lower cutoff $x_<$ for $x_> = 1$. We have set $N = 10$.

reasoning: the breaking criterion, Eq. (4.12), determines which fiber will fail next. If we are to be guaranteed that one of the two edge fibers will fail next, we need that

$$\frac{\left(\frac{3}{2}\right)}{1} > \frac{1}{x_<}, \tag{4.35}$$

since the left-hand side is the smallest possible ratio that the two edge fibers may attain and the right-hand side is the largest possible ratio that the other fibers may attain. This inequality may be rewritten

$$\frac{3}{2} \quad x_< > 1, \tag{4.36}$$

We are guaranteed that one of the two fibers adjacent to the one that failed first will fail. Clearly, the inequality secures that only fibers at the edge of the single existing hole will continue to fail.

We will discuss this phenomenon in detail. However, before doing this, we need to define the tools that we will use when going beyond the failure of the two first fibers. The most important concept in the following is the *hole size distribution* $p_h(m, k)$.

4.1.5 Hole Size Distribution

We have just calculated the probability to find a hole of size 2 when two fibers have failed, $p_h(2, 2)$. We now generalize $p_h(2, 2)$ to the probability to find a hole of size m when k fibers have broken, $p_h(m, k)$.

Perhaps the most important lesson from the calculation of $p_h(2, 2)$ is that even very early in the failure process, the force redistribution plays a significant role in the evolution of the model. As soon as the first fiber fails, the local-load-sharing model evolves differently from the equal-load-sharing model: compare Eq. (4.21) with Eq. (4.40).

A second lesson is that finding $p_h(m, k)$ is exceedingly difficult except for the simplest nontrivial case, that is, for $k = 2$. In this section, we determine $p_h(m, k)$ for $k > 2$. We need to do this numerically.

However, first we need to define precisely the hole size distribution.

4.1.5.1 Defining the Hole Size Distribution

How would one characterize the distribution of hole sizes by using the computer? We would generate sample after sample, and, for each count the number of times we encounter a hole of size $1, 2, \ldots m$, up to k when k fibers have failed. We accumulate these numbers in an array $H(m, k)$. After we are done with all the samples, we divide each entry in $H(m, k)$ by the number of samples. Hence, $H(m, k)$ then contains the average number of holes of size m per sample when k fibers in total have failed.

We must have that

$$\sum_{m=1}^{k} mH(m,k) = k. \tag{4.37}$$

We divide by k and get

$$\sum_{m=1}^{k} \frac{m}{k} \quad H(m,k) = 1, \tag{4.38}$$

allowing us to define

$$p_{\mathrm{h}}(m,k) = \frac{m}{k} \quad H(m,k). \tag{4.39}$$

This is the hole size distribution.

4.1.5.2 Hole Size Distribution for Equal Load Sharing

In order to gain some intuition with respect to $p_{\mathrm{h}}(m,k)$, let us calculate $p_{\mathrm{h}}(m,3)$ for the *equal-load-sharing model.* In this case, all fibers carry the same force at any stage in the process. Hence, only the thresholds determine which fibers will fail next. We may, therefore, use simple combinatorics to determine $p_{\mathrm{h}}(m,3)$.

The probability that there is a hole of size 2 when $k = 2$ is

$$p_{\mathrm{h}}(2,2) = \frac{2}{N-1}. \tag{4.40}$$

This is just the probability that the fiber with the smallest remaining threshold after the first fiber failed is adjacent to it. The probability that a fiber adjacent to these two failed fibers also fails is $p_{\mathrm{h}}(2,2)[2/(N-2)]$. However, this is not the only way a hole of size 3 may form. We may also have a configuration for $k = 2$ where two next-nearest neighbor fibers have failed. The probability for this configuration is also $2/(N-1)$. If the fiber between these two fibers fails next, we have a hole of size 3. The probability for this event is $1/(N-2)$. We combine these results and have[1)]

$$p_{\mathrm{h}}(3,3) = p_3 = \frac{6}{(N-1)(N-2)}, \tag{4.41}$$

where we have introduced the quantity p_3. This is the probability that we create a single hole of size 3.

If we do not create a single hole of size 3 when $k = 3$, we either create a hole of size 2 and a hole of size 1 or three holes of size 1. The probability for the first is p_{2+1}, and the probability for the second is p_{1+1+1} in obvious notation.

Using reasoning along the same lines as that leading to Eq. (4.41), we find

$$p_{2+1} = \frac{6(N-4)}{(N-1)(N-2)}, \tag{4.42}$$

1) This expression is easily generalized to any k,

$$p_{\mathrm{h}}(k,k) = \frac{k!}{(N-1)\cdots(N-k+1)}.$$

and

$$p_{1+1+1} = \frac{(N-4)(N-5)}{(N-1)(N-2)}. \tag{4.43}$$

We see that $p_3 + p_{2+1} + p_{1+1+1} = 1$, as it should be.

We may now construct $p_{\rm h}(m,3)$. Since p_{1+1+1} contributes three holes of size 1 and p_{2+1} one hole of size 1, we have

$$p_{\rm h}(1,3) = \frac{1}{3} \quad (3p_{1+1+1} + p_{2+1}) = \frac{(N-3)(N-4)}{(N-1)(N-2)}. \tag{4.44}$$

Likewise, p_{2+1} contributes one hole of size 2, and therefore, we have

$$p_{\rm h}(2,3) = \frac{2}{3} \quad p_{2+1} = \frac{4(N-4)}{(N-1)(N-2)}. \tag{4.45}$$

We have $p_{\rm h}(1,3) + p_{\rm h}(2,3) + p_{\rm h}(3,3) = 1$.

We note from the previous equations that

$$\frac{p_{\rm h}(m+1,3)}{p_{\rm h}(m,3)} \sim \mathcal{O}\left(\frac{1}{N}\right), \tag{4.46}$$

This carries over to arbitrary k. A very crude argument can be used in the limit of large N to demonstrate this. If we suppose that the probability that a fiber fails is $\tilde{P} = k/N$ – which is a valid assumption since the forces play no role in selecting the fibers for failure, the probability that there are m failed fibers in a row is $(1-\tilde{P})^2\tilde{P}^m$ in the limit we assume. Hence, the number of holes of size m is $H(m,k) = \mathcal{N}\tilde{P}^m$ where $\mathcal{N}$ is a normalization. Using Eq. (4.39), the hole size distribution may then be approximated by

$$p_{\rm h}(m,k) = \frac{m(1-\tilde{P})^2\tilde{P}^{m-1}}{1-[1+k(1-\tilde{P})]\tilde{P}^k}, \tag{4.47}$$

where we have used the normalization condition $\sum_{m=1}^{k} p_{\rm h}(m,k) = 1$. We illustrate how well this approximation works in Figure 4.10 when both k and N are large (in this case $k = 200$ and $N = 1000$).

On the other hand, Eq. (4.47) predicts

$$p_{\rm h}(3,3) = \frac{27}{27 + N(N+6)}, \tag{4.48}$$

for $m = k = 3$. This should be compared with the exact expression (4.41). For small k, the approximation is not very precise.

4.1.5.3 Hole Size Distribution for Local Load Sharing: Localization

Compare now the hole size distribution for the local-load-sharing model with that of the equal-load-sharing model, Figure 4.11 versus 4.10. In the local-load-sharing case, there is a behavior not unlike that of the equal-load-sharing model for small m. For large m, on the other hand, there is a broad maximum.

The reason for this broad maximum in the local-load-sharing case is the following: early in the breakdown process, the failures proceed in a manner not unlike that of the equal-load-sharing model. However, at some point, one of the holes

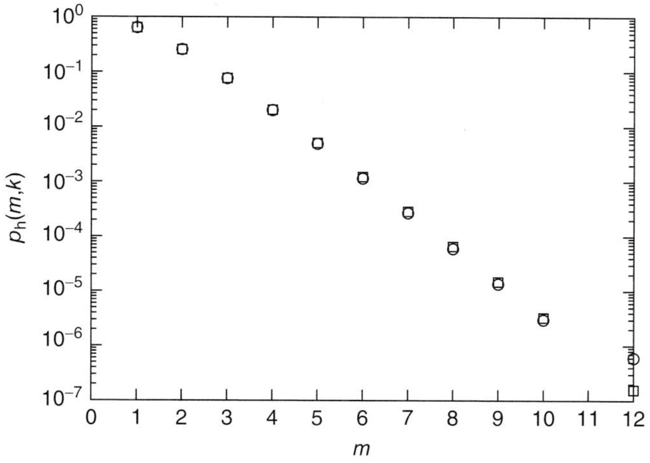

Figure 4.10 Simulation results for $p_h(m, k)$ for the equal-load-sharing model for $N = 1000$ and $k = 200$ based on 10^5 samples (circles). Equation (4.47) is plotted as squares.

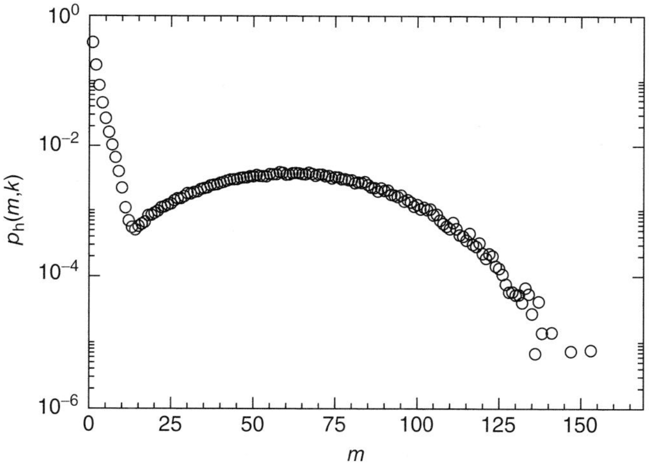

Figure 4.11 $p_h(m, k)$ for the local-load-sharing model for $N = 1000$ and $k = 200$ based on 10^5 samples. The threshold distribution was uniform on the unit interval.

starts growing, freezing all other failure activity in the model. Only the fibers at the edge of the growing hole fail.

This is very reminiscent of the localization phenomenon we met in Section 4.1.4.2. However, there is a big difference: in Figure 4.11, the threshold distribution was the uniform one on the unit interval, and hence, *there was no gap* $x_< > 0$ in the threshold distribution! The localization effect we saw in Section 4.1.4.2 was based on the presence of a gap, see Eq. (4.36). However, as the failure process proceeds, fibers are removed. These will have the smallest thresholds as long as they are not adjacent to holes. As these smallest thresholds

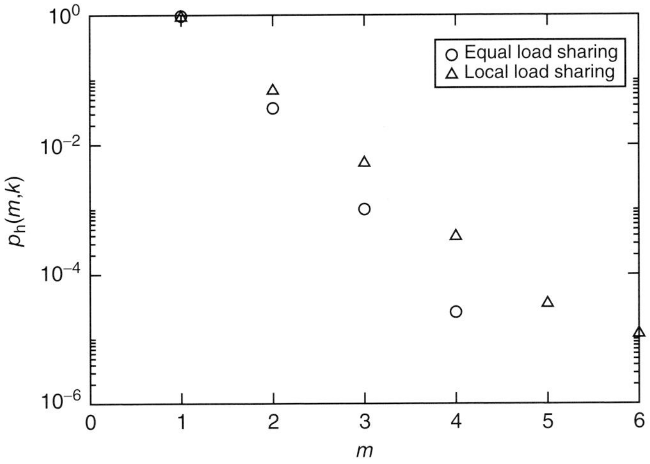

Figure 4.12 $p_h(m, k)$ for the local-load-sharing model and equal-load- sharing model for $N = 1000$ and $k = 20$ based on 10^5 samples. The threshold distribution was uniform on the unit interval.

are removed, the remaining fibers will effectively see a threshold distribution with a nonzero gap. Hence, localization sets in.

We will investigate this mechanism in the following. However, before delving into this, we compare $p_h(m, k)$ for equal and local load sharing when k is small. This is shown in Figure 4.12. The hole size distribution falls off more slowly for local load sharing than for equal load sharing. We already saw this for $p_h(2, 2)$: for the uniform threshold distribution on the unit interval, $p_h(2, 2) = 4/N$ for local load sharing, whereas it is $p_h(2, 2) = 2/N$ for equal load sharing – both examples taken in the limit of large N.

4.1.6
Estimating the Strength of the Local-Load-Sharing Model

In Section 4.1.3, we considered the strength σ_c of the local-load-sharing fiber bundle as a function of N. Figure 4.4 shows numerically that

$$\sigma_c^{-1} = A \frac{\ln(N^2)}{\ln(\ln(N^2))} + B, \tag{4.49}$$

see Eq. (4.13). It is the aim of this section to provide an approximate derivation on this result – approximative since an exact calculation would be based on the knowledge of $p_h(m, k)$ beyond $k = 2$, which is a *fontes Nili querere,* an impossible task.

We will approach the problem as follows: we will (1) determine a lower bound for the typical size of the largest hole after k fibers have failed. (2) We will then estimate for which k the largest hole has become so large that localization sets in. At this point, the largest hole will grow by having its edge fibers failing successively. The stress at which localization sets in is the critical stress σ_c.

4.1.6.1 Lower Bound for the Largest Hole

Rather than using the $p_h(m, k)$ for local-load-sharing model, which we do not know, we use the hole size distribution for the equal-load-sharing model in the limit of large N, Eq. (4.47). The corresponding cumulative probability is

$$P_h(m, k) = \frac{1 - [m(1 - \tilde{P}) + 1]\tilde{P}^m}{1 - [k(1 - \tilde{P}) + 1]\tilde{P}^k}. \tag{4.50}$$

This substitution will underestimate the probability $p_h(m, k)$, see Figure 4.12.

Let us now focus on how a given sample fails. When k fibers have failed, let there be h holes in the array of fibers. We number them from 1 to h. Hole number j has size m_j, and we have

$$\sum_{j=1}^{h} m_j = k. \tag{4.51}$$

The cumulative hole probability, Eq. (4.50), may be simplified when taking into account that $kp^k = k(k/N)^k \ll 1$ when $N \gg k$ and $k \gg 1$. The cumulative hole size probability then simplifies into

$$P_h(m, k) = 1 - [m(1 - \tilde{P}) + 1]\tilde{P}^m. \tag{4.52}$$

If we now assume $m \gg 1$, it simplifies further,

$$P_h(m, k) = 1 - m(1 - \tilde{P})\tilde{P}^m. \tag{4.53}$$

We note that

$$m(1 - \tilde{P}) < k \tag{4.54}$$

for all m. Hence, we have

$$P_h(m, k) > 1 - k\tilde{P}^m, \tag{4.55}$$

for all m. Since we wish to establish a lower bound for the size of the largest hole, we may replace the hole size probability (4.52) by

$$P_h(m, k) = 1 - k\tilde{P}^m. \tag{4.56}$$

We remind the reader that $\tilde{P} = k/N$.

We order the hole sizes, $m_1, m_2, \cdots, m_h \to m_1, m_2, \cdots, m_h$, where $m_1 \leq m_2 \leq \cdots \leq m_h$. We now ask the following question: what is the average size of the jth largest hole, $\langle m_j \rangle$, where the average is taken over an ensemble of samples? *Order statistics* (Appendix B.2.2) gives us the sample average of $\langle m_j \rangle$ through Eq. (B.28),

$$P_h(\langle m_j \rangle, k) = \frac{j}{h + 1}. \tag{4.57}$$

The trouble with using this result in the following is that we do not know h. We are *not* generating a fixed number of holes, h. Rather, we break a fixed number of fibers, k, and h will vary from sample to sample (but in practice, not so much). Hence, we will have to replace h by $\langle h \rangle$.

We now determine $\langle m_j \rangle$ and $\langle h \rangle$ as a function of k. We base the derivation on the constraint (4.51).

From Eqs. (4.56) and (4.57), we have

$$\langle m_j \rangle = \frac{1}{\ln(\tilde{P})} \ln\left(1 - \frac{j}{\langle h \rangle + 1}\right) - \frac{\ln(k)}{\ln(\tilde{P})}. \tag{4.58}$$

In the following, we denote $\langle m_{(j)} \rangle$ as $m_{(j)}$ and $\langle h \rangle$ as h in order not to make the notation too cumbersome.

Combining Eq. (4.58) with Eq. (4.51), we find

$$k = \sum_{j=1}^{h} m_j = \frac{\ln\left[\prod_{j=1}^{h}\left(1 - \frac{j}{h+1}\right)\right]}{\ln(\tilde{P})} - h\,\frac{\ln(k)}{\ln(\tilde{P})} = \frac{\ln\left[\frac{h!}{(h+1)^h k^h}\right]}{\ln(\tilde{P})}. \tag{4.59}$$

We now need to solve this equation with respect to h, finding $h = h(k)$. Assuming $h \gg 1$ and invoking the Stirling approximation

$$h! = \sqrt{2\pi h}\left(\frac{h}{e}\right)^h, \tag{4.60}$$

Eq. (4.59) is transformed into

$$k = \frac{\ln\left[\sqrt{2\pi h}\left(\frac{h}{(h+1)ke}\right)^h\right]}{\ln(\tilde{P})}. \tag{4.61}$$

By neglecting terms of order $\ln(h)$ or smaller compared to h and assuming $\ln(k) \gg 1$, we find

$$h = -\left(\frac{\ln(\tilde{P})}{\ln(k)}\right) k = \left(\frac{\ln(N)}{\ln(k)} - 1\right) k = \lambda k, \tag{4.62}$$

where we have defined

$$\lambda = -\frac{\ln(\tilde{P})}{\ln(k)} = \left(\frac{\ln(N)}{\ln(k)}\right) - 1. \tag{4.63}$$

From Eq. (4.58), we find that the largest hole to be expected when k fibers have failed is

$$m_h = \frac{\ln\left(\frac{1}{(h+1)k}\right)}{\ln(\tilde{P})} \approx \frac{\ln\left(\frac{1}{hk}\right)}{\ln(\tilde{P})} = \frac{\ln(\lambda k^2)}{\ln(N/k)}, \tag{4.64}$$

where we have assumed that $h \gg 1$.

Using Eq. (4.64), we now calculate the load on each of the two nearest neighbors to the largest hole (assuming that they are not also bordering on a second hole), finding

$$\sigma_{\mathrm{h}}(x) = \frac{m_h}{2} \quad x = \frac{x\ln(\lambda k^2)}{2\ln(N/k)}. \tag{4.65}$$

We will now combine these results with the failure criterion, Eq. (4.12), described in Section 4.1.2.

4.1.6.2 Competing Failure Mechanisms

k fibers have failed. How do we decide upon which fiber will be number $k + 1$ to break? Equation (4.12),

$$\max_i \left(\frac{\sigma_i(1)}{x_i} \right) = \frac{1}{x},$$

tells us to identify the fiber that has the largest ratio of force over threshold when the elongation x is unity.

From the failure criterion (4.12), we see that there are two competing mechanisms at play when the fibers fail: either a fiber can fail because its threshold is small, that is, the ratio $\sigma_i(1)/x_i$ is large since x_i is small, or it can fail because the force it carries, $\sigma_i(1)$, is large.

In Section 4.1.4, we calculated $p_\mathrm{h}(2, 2)$ exactly. For the uniform threshold distribution on the unit interval, we found that $p_\mathrm{h}(2, 2) \to 4/N$ as $N \to \infty$, whereas it was $2/N$ in the equal-load-sharing model as $N \to \infty$. Hence, there is an effect of the redistribution of the stresses from the very beginning of the failure process. However, the probability that a fiber that is not bordering the hole created by the first failed fiber will fail is $1 - 4/N \approx 1$ and $1 - 2/N \approx 1$ in the local-load-sharing and the equal-load-sharing models, respectively. Fibers fail because of their small thresholds, not because they are highly stressed early in the breakdown process.

Later on in the breakdown process, localization sets in. At this point, fibers fail because they are being highly stressed.

We now make the following approximation: we divide the failure process into two parts: (1) up to a critical $k = k_\mathrm{c}$, the failure due to small thresholds dominates the process and (2) for larger $k > k_\mathrm{c}$, the stresses dominate. At k_c, the two types of processes are in balance.

We determine k_c by equating the ratio between force at $x = 1$ and the threshold for the fibers not bordering holes with the corresponding ratio for the two fibers bordering the largest hole,

$$\max_{i'} \left(\frac{\sigma_{i'}(1)}{x_{i'}} \right) = \frac{\sigma_\mathrm{h}(1)}{x_\mathrm{h}}, \tag{4.66}$$

where the index i' runs over all fibers not bordering any hole. We have that $\sigma_{i'}(1) = 1$ since these fibers are not bordering any holes. x_h is the smallest threshold bordering on the largest hole. We replace it by the average threshold value $\langle x \rangle$. Equation (4.66) then becomes

$$\max_{i'} \left(\frac{1}{x_{i'}} \right) = \frac{\sigma_\mathrm{h}(1)}{\langle x \rangle}. \tag{4.67}$$

The thresholds are ordered, $x_1 \le x_2 \le \cdots \le x_N$. When $k < k_\mathrm{c}$ fibers have failed, we have that

$$\max_{i'} \left(\frac{1}{x_{i'}} \right) \approx \frac{1}{x_k}, \tag{4.68}$$

where we are assuming that x_k is not situated next to a hole.

From order statistics (Appendix B.2.2), we have that (B.28)

$$P(\langle x_k \rangle) = \frac{1}{N+1} \approx \frac{k}{N} = \tilde{P}, \tag{4.69}$$

where $P(x)$ is the cumulative threshold probability. We combine this expression with Eq. (4.67) to find

$$\langle x_{k_c} \rangle = P^{-1}\left(\frac{k_c}{N}\right) = \frac{\langle x \rangle}{\sigma_h(1)}. \tag{4.70}$$

We now use our estimate (4.65) for $\sigma_h(1)$, and Eq. (4.70) becomes

$$\frac{k_c}{N} = P(\langle x_{(k_c)} \rangle) = P\left(\frac{\langle x \rangle}{\sigma_h(1)}\right) = P\left(\frac{2\langle x \rangle \lambda_c \ln(k_c)}{\ln(\lambda_c k_c^2)}\right), \tag{4.71}$$

where $\lambda_c = \ln(N/k_c)/\ln(k_c)$. Equation (4.71) gives k_c as a function of N.

When the failure process reaches k_c, it localizes to the edge of the largest hole. From this point on, it is unstable, that is, $x_{k_c} > x_{k_c+1} > x_{k_c+2} > \cdots > x_N$. Hence,

$$x_c = x_{k_c}. \tag{4.72}$$

The maximum force that the fiber bundle can sustain without failing is then given by

$$F_c = N x_c = N \sigma_c, \tag{4.73}$$

since the model is constructed in such a way that the total force carried by the surviving fibers is always equal to the force that would have been carried by all the fibers had they all been intact; see Eq. (4.7).

4.1.6.3 Uniform Threshold Distribution

In the following, we assume $P(x) = x$ for $0 \le x \le 1$. Hence, $\langle x \rangle = 1/2$. Equation (4.71) can then be written

$$\left(\frac{N}{k_c}\right) \ln\left(\frac{N}{k_c}\right) = \ln(\lambda_c k_c^2), \tag{4.74}$$

which after some manipulations may be rewritten as

$$\left(\frac{N}{k_c} + 2\right) \ln\left(\frac{N}{k_c}\right) = \ln(\lambda_c) + \ln(N^2). \tag{4.75}$$

If we now assume that $N/k_c \gg 2$ and $\ln(\lambda_c) \ll \ln(N^2)$, this equation may be simplified,

$$\left(\frac{N}{k_c}\right) \ln\left(\frac{N}{k_c}\right) = \ln(N^2). \tag{4.76}$$

This is of course a transcendental equation in k_c. However, we may approximate the solution by noting that

$$f(t) \ln[f(t)] = t \tag{4.77}$$

has as an approximate solution

$$f(t) \approx \frac{t}{\ln(t)}. \tag{4.78}$$

By reinserting this result back into Eq. (4.77), we find that the error is of order $t \ln(\ln(t)) / \ln(t)$. Equation (4.76) may thus be solved approximately as

$$k_c = \frac{N}{\ln(N^2)} \quad \ln(\ln(N^2)), \tag{4.79}$$

with an error of the order of $\ln(N^2)\ln(\ln(\ln(N^2)))/\ln(\ln(N^2))$.

The approximative critical stress is then

$$\frac{1}{\sigma_c} = \frac{\ln(N^2)}{\ln(\ln(N^2))}, \tag{4.80}$$

where we have used that

$$\sigma_c = x_c = \frac{k_c}{N}. \tag{4.81}$$

We now compare these considerations to results from direct numerical simulation. Zhang and Ding [33] measured the critical extension numerically and proposed the fit

$$\frac{1}{\sigma_c} = a \ln(N) + b, \tag{4.82}$$

where $a = 0.70$ and $b = 2.31$ for $5 < \ln N < 13$ ($N = 148$ to $442\,413$). The best fit to our data of the functional form (4.82) we obtain for $a = 0.70$ and $b = 2.60$. We show the data and the fit in Figure 4.13. We have also plotted our estimate for σ_c, Eq. (4.79), in the same figure.

However, inspired by Eq. (4.79), we propose as a replacement for the functional form for $1/\sigma_c$ proposed by Zhang and Ding (4.82), the form

$$\frac{1}{x_c} = A\frac{\ln(N^2)}{\ln(\ln(N^2))} + B. \tag{4.83}$$

Setting $A = 1.6$ and $B = -1.04$, this expression fits the data very well as seen in Figure 4.13, in fact much better than Eq. (4.82) does.

We have thus in this section shown that an approximate critical stress σ_c in the local-load-sharing model approaches zero as $\ln(\ln(N))/\ln(N)$ as $N \to \infty$ for the uniform threshold distribution.[2] From numerical simulations, we find that this is also the way the σ_c of the local-load-sharing model itself approaches zero as $N \to \infty$.

2) Remember that $\ln(\ln(N^2)) \cong \ln(\ln(N))$ when $N \to \infty$.

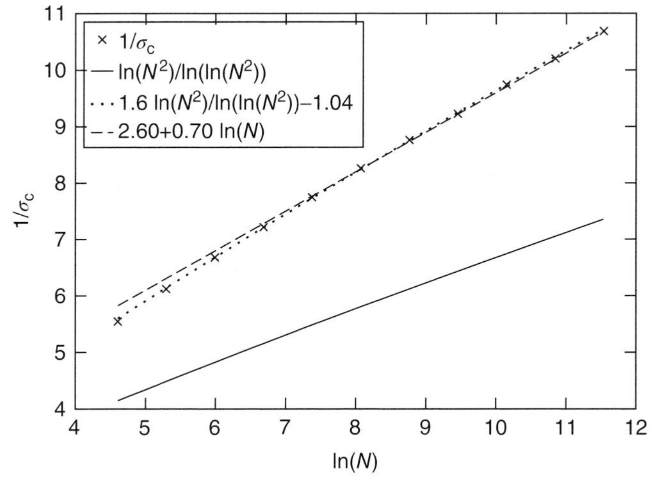

Figure 4.13 Inverse critical stress $1/\sigma_c$ versus $\ln(N)$ based on 10^5 samples for $N = 100$ to 2000 samples for $N = 102\,400$ (crosses). We also show the predictions of Eqs. (4.82) (broken curve) and (4.83) (dotted curve). The derived approximative solution (4.80) is also shown (black solid line).

4.1.7
Force and Elongation Characteristics

The force versus elongation characteristics of the local-load-sharing model was given in Eq. (4.7),

$$F = Nx.$$

This should be compared to for example, the force-elongation characteristics of the equal-load-sharing model

$$F = N(1 - P(x))x,$$

where $P(x)$ is the cumulative threshold distribution.

On the other hand, force or elongation plotted against the number of failed fibers shows nontrivial behavior in the local-load-sharing model. We show σ versus k for the flat threshold distribution on the unit interval in Figure 4.14. We have rescaled $x \to x[A\ln(N^2)/\ln(\ln(N^2)) + B]$ and $k \to k[A\ln(N^2)/\ln(\ln(N^2)) + B]/N$ as suggested by Eq. (4.83). There is, as expected, data collapse right up to the maximum value (k_c, x_c). For larger k values, there is no data collapse. Why?

We investigate this question in Figure 4.15. Here we have plotted $1/\sigma$ versus k. When the critical value (k_c, x_c) is reached, the fiber bundle undergoes catastrophic failure. This is due to localization as described in Section 4.1.6.2. When localization sets in, we expect the characteristics to follow the equation

$$\frac{1}{\sigma} = \frac{m_h + k - k_c}{2\langle x_i \rangle}, \tag{4.84}$$

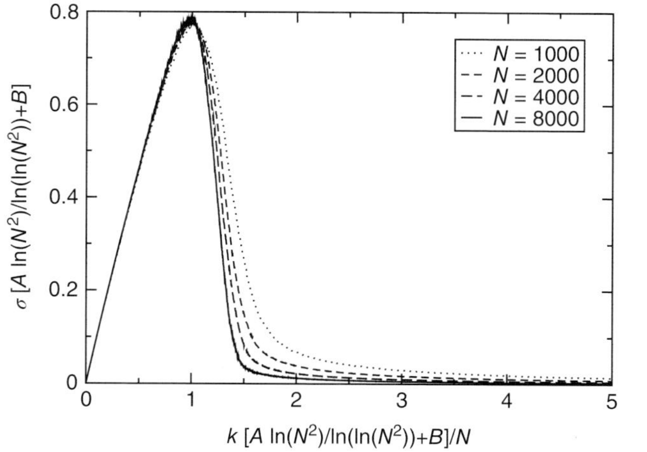

Figure 4.14 $\sigma(A \ln(N^2)/\ln(\ln(N^2)) + B)$ versus $k(A \ln(N^2)/\ln(\ln(N^2)) + B)/N$ with the values given in Figure 4.10, $A = 1.6$ and $B = -1.04$. The threshold distribution was uniform on the unit interval. Each curve is based on 2000 samples.

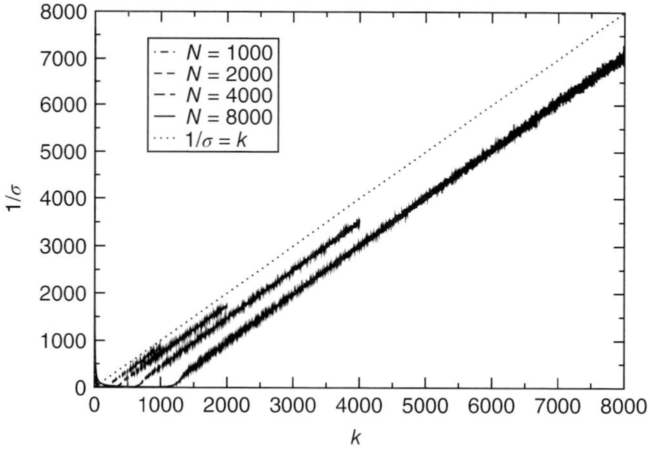

Figure 4.15 $1/\sigma$ versus k based on the uniform threshold distribution on the unit interval. The straight line signifies localization and we expect it to follow $k/2\langle x_i \rangle$ with $\langle x_i \rangle = 1/2$. The figure is based on 2000 samples for each N value.

or

$$\frac{1}{\sigma} = \frac{k}{2\langle x_i \rangle} \tag{4.85}$$

for large k. Our reasoning is as follows: once localization sets in for $k \geq k_c$, the force at the edge of the largest hole is so large that the fibers there fail. They carry a stress $[m_h + (k - k_c)]/2$, where m_h is the size of the largest hole when localization sets in, see Eq. (4.65). We see in Figure 4.15 that the unstable part of the characteristics indeed follows the functional form of Eq. (4.85).

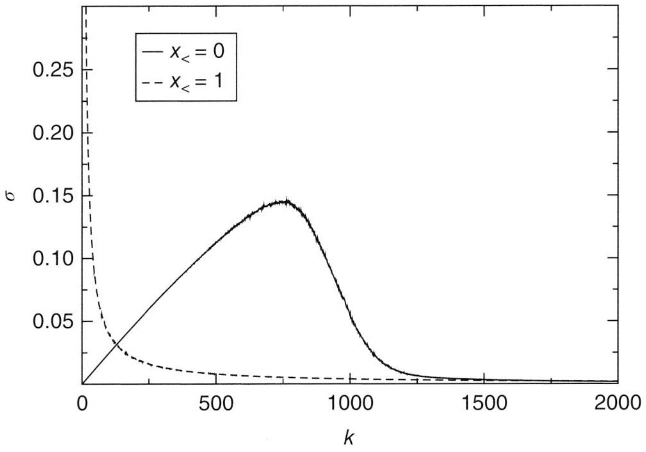

Figure 4.16 σ versus k based on the threshold distribution $P(x) = 1 - \exp\,[x_< - x]$ with $x_< \leq x < \infty$. We have set $x_< = 0$ and 1. The figure is based on 2000 samples with $N = 4000$.

We show in Figure 4.16 σ versus k for the exponential threshold distribution $P(x) = 1 - \exp\,(x_< - x)$ for $x_< \leq x < \infty$. We show two cases, $x_< = 0$ and $x_< = 1$. In Section 4.1.4.2, we discussed localization using this threshold distribution as an example. The probability that the failure of one of the two fibers neighboring the fiber that first failed (for $k = 1$) is given by Eq. (4.32). For $x_< = 0$, it is $4/N$ for large N whereas it is $1 - 1/e \approx 0.63$ for $x_< = 1$. In the latter case, the fracture process is essentially localized from the outset. The difference between $x_< = 0$ and $x_< = 1$ is clearly seen in Figure 4.16.

In Figure 4.17 we plot $1/\sigma$ versus k in analogy with Figure 4.16 for the cumulative threshold probability $P(x) = 1 - \exp\,[x_< - x]$ with $x_< \leq x < \infty$. We have set $x_< =$ 0 and 1, respectively. $\langle x_i \rangle$ is then 1 and 2 for the two cases. We see that also in this case, Eq. (4.85) is followed when the failure process is localized. In particular, we see this behavior from the very start of the failure process for $x_< = 1$ – the case for which localization sets in almost immediately.

4.1.8
Burst Distribution

The relation between force F and elongation x is, as pointed out several times already, simple in the local-load-sharing model. Equation (4.7),

$$F = Nx,$$

states that the force is always proportional to the elongation. Hence, bursts in F are then in one-to-one correspondence with bursts in x. We, therefore, concentrate on the latter type of bursts.

The first to study the burst distribution in the local-load-sharing model was Hansen and Hemmer [34]. Studying only the uniform threshold distribution on

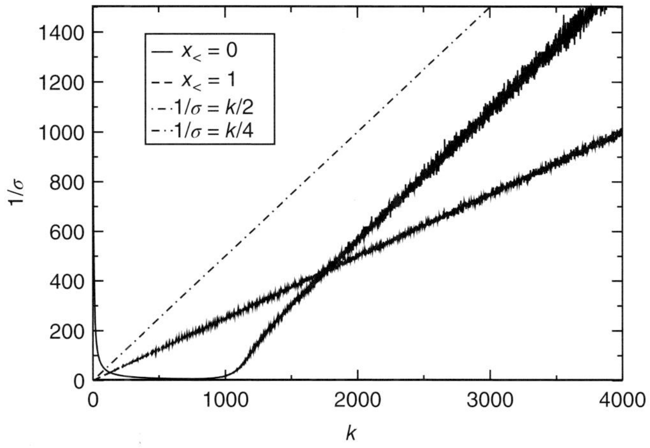

Figure 4.17 $1/\sigma$ versus k based on the threshold distribution $P(x) = 1 - \exp[x_< - x]$ with $x_< \leq x < \infty$. We have set $x_< = 0$ and 1. The two straight lines are $1/\sigma = k/2\langle x\rangle$ with $\langle x\rangle = 1$ ($x_< = 0$) and $\langle x_i\rangle = 2$ ($x_< = 1$), respectively. The figure is based on 2000 samples with $N =$ 4000.

the unit interval, they reported a burst distribution

$$D(\Delta) \propto \Delta^{-\xi} \tag{4.86}$$

with $\xi = 4.5$. Shortly afterward, Zhang and Ding [33] found that the burst exponent ξ depends on the threshold distribution and on the number of fibers in the bundle, N. They investigated cumulative threshold distributions of the form $P(x) = x^\beta$ where $\beta = 1$ and 2 and $0 \leq x \leq 1$. We show a corresponding plot for $\beta = 1, 1/2, 1/5, 1/10$, and 1/25 in Figure 4.18.

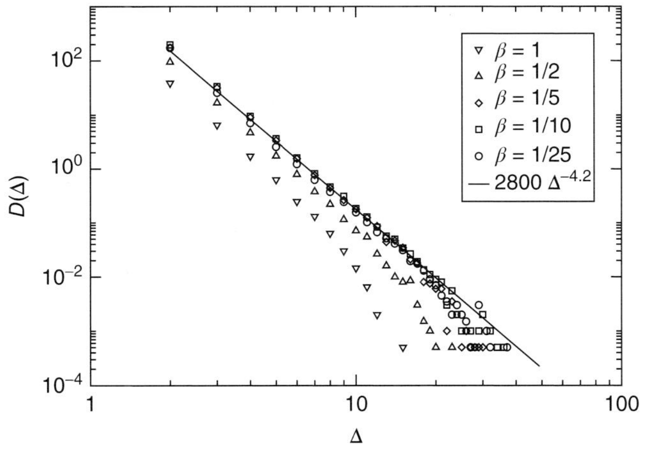

Figure 4.18 Burst distribution $D(\Delta)$ for the threshold distributions $P(x) = x^\beta$ with $0 \leq x \leq 1$. The figure is based on 20 000 samples with $N = 10\,000$ for each value of β.

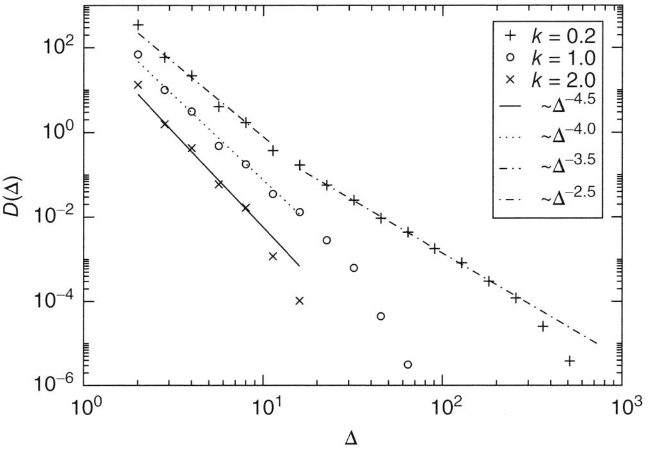

Figure 4.19 Burst distribution $D(\Delta)$ for the threshold distributions $P(x) = 1 - \exp[-x^k]$ with $0 \leq x < \infty$ and $k = 0.2$, 1 and 2. The figure is based on 20 000 samples of size $N = 10\,000$ for each data set.

In 1997, Kloster *et al.* [18] managed to show analytically that there is indeed no universal power law describing the burst distribution in the local-load-sharing model when the threshold distribution is uniform. Rather, there is an effective exponent ξ, which depends on N – just as observed by Zhang and Ding [33]. We hesitate to reproduce the derivation of this result by Kloster *et al.* here. Rather, we direct the interested reader to the original paper.

In Figure 4.19, we show the burst distribution for a very different class of cumulative threshold distributions, namely $P(x) = 1 - \exp(-x^k)$, where $0 \leq x < \infty$. We see that for the chosen values of k, there is no universal burst exponent ξ. However, for $k = 0.2$, there is a crossover for bursts larger than $\Delta \approx 20$ to a different distribution. The burst exponent here is essentially 2.5, the value found in the equal-load-sharing model!

With this remark, we now turn to the *inclusive bursts* – that is also counting all bursts within bursts, see Section 3.3.3. We show their distribution in Figure 4.20 for the cumulative threshold distribution $P(x) = 1 - \exp\ (x_< - x)$ with $x_< = 1$ and $x_< \leq x < \infty$. With this value, we know that localization sets in from the very beginning of the breakdown process. There is only one growing hole.

The localized behavior makes the inclusive burst distribution simple to derive. Suppose that k fibers have failed. The last failure happened at $x = x_k$. An inclusive burst of size Δ would then happen with probability $P(x_k)^{\Delta-1}[1 - P(x_k)]$ as the next $\Delta - 1$ fibers would have thresholds less than x_k, whereas the fiber after that would need a threshold larger than x_k. The probability density that fiber k has a threshold x_k is $p(x_k) = \mathrm{d}P(x_k)/\mathrm{d}x_k$. Hence, the inclusive burst distribution is

$$\frac{D_i(\Delta)}{N} = \int_0^\infty \mathrm{d}x \quad p(x)P(x)^{\Delta-1}[1 - P(x)] = \frac{1}{\Delta + \Delta^2}. \tag{4.87}$$

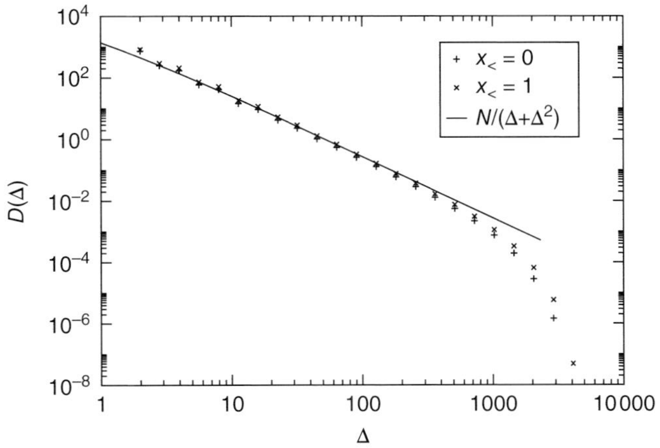

Figure 4.20 Inclusive burst distribution $D(\Delta)$ in the local-load-sharing model for the threshold distributions $P(x) = 1 - \exp(x_< - x)$ with $x_< \leq x < \infty$. The curve $N/(\Delta + \Delta^2)$ fits the data very well. The figure is based on 20 000 samples when $N = 10\ 000$.

We plot this simple result together with the inclusive burst distribution in Figure 4.20. Note that there are no adjustable parameters in this expression.

More cannot be said here about bursts in the local-load-sharing model. Clearly, it is an unfinished chapter. But, the reader should rejoice at that: there is more to do!

4.2 Local Load Sharing in Two and More Dimensions

So far, we have considered the local-load-sharing model in one dimension, where the fibers are placed along a line. What happens when we implement it in two dimensions where the fibers may be placed at the nodes of some lattice?

First, let us think through what happens in the equal-load-sharing model when going from one dimension to two dimensions. Of course, nothing happens. In the equal-load-sharing model, the geometrical placement of the fibers does not matter. The clamps that hold the fibers are both infinitely stiff and all forces are transmitted equally between all surviving fibers no matter where they are.

In the local-load-sharing model, on the other hand, the placement of the fibers matters. This increases the complexity of the model tremendously. Whereas in one dimension, each hole created by the failed fibers would be flanked by two surviving fibers, in two dimensions, holes can have very irregular shapes. In spite of this, as we shall see, we will be able to say quite a lot about the behavior of the model.

Before delving into the details, let us state explicitly the definition of the local-load-sharing model in two dimensions: all the fibers next to a hole of failed fibers will absorb equally the force that the failed fibers in the hole would have carried were they intact. Hence, Eq. (4.7),

$$F = Nx,$$

holds. If a fiber forms a bridge between two or more holes, the force it carries is the sum of the forces from each of the holes.

Some earlier work has been performed on the two-dimensional local-load-sharing model, see for example, [15, 35]. In the following we base our discussion on Sinha *et al.* [36].

4.2.1 Localization

We show in Figure 4.21 the model seen from "above." That is, we are seeing each fiber along its axis. We represent surviving fibers as black dots and failed fibers as white dots. In the figure, we see the situation when $k = 5120$ fibers have failed. The cumulative threshold distribution was $P(x) = 1 - \exp\,[x_< - x]$ where $x_< \leq x < \infty$ with $x_< = 0$ in the left panel and $x_< = 1$ in the right panel.

Clearly, there is a huge difference between the two panels in Figure 4.21. In the left panel, the threshold distribution goes all the way to zero strength, whereas in the right panel, there is a lower cutoff equal to one. In Section 4.1.4.2, we demonstrated that a sufficient large nonzero lower cutoff in the threshold distribution quickly leads to localization in the one-dimensional local-load-sharing model. This also happens here as is evident from the figure.

As we have seen, when localization sets in in one dimension, the fiber bundle fails due to a single growing hole. The structure is much richer in two dimensions. Whereas there are only two fibers that form the edge of the unstable hole, the two-dimensional structure is extremely convoluted as seen in the right panel in Figure 4.21. If the reader thinks "fractal" at this point, he or she is correct. In fact,

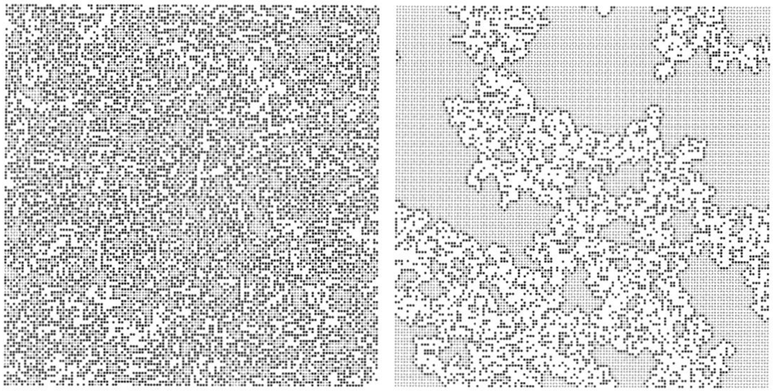

Figure 4.21 Here we see the two-dimensional local-load-sharing model from "above". Each intact fiber is shown as a black dot and each failed fiber as a white dot. The system size is $N = 128^2$. We show the model when $k = 5120$ fibers have failed. The cumulative threshold distribution was $P(x) = 1 - \exp\,[x_< - x]$ where $x_< \leq x < \infty$. In the left panel, we have $x_< = 0$ and in the right panel $x_< = 1$.

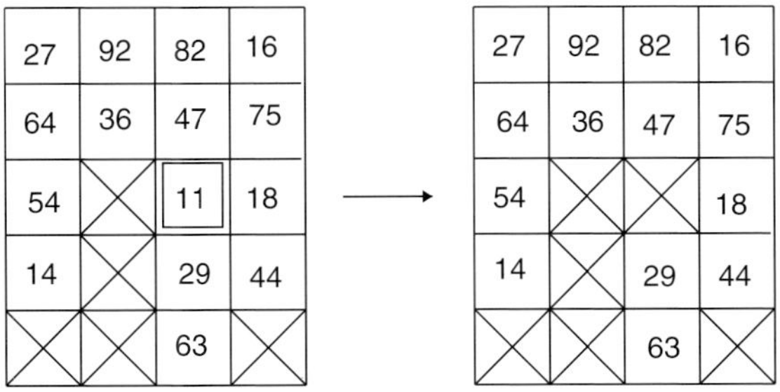

Figure 4.22 The invasion percolation model: a random number – here an integer between 0 and 100 – is assigned to each square in the tiling. We then invade the tiling from below, always choosing the tile with the smallest random number assigned to it, which is next to the already invaded tiles. We illustrate the process after five tiles have been invaded. We have marked the tile with the smallest random number next to the invaded tiles. At the next step, this tile is invaded as shown in the right panel.

the structure of the front is well known and has been the subject of intense study since the 1980s. It turns out when there is localization that the local-load-sharing model in two dimensions is equivalent to the *invasion percolation model.*

We show in Figure 4.22 a square tiling. Each tile has been given a random integer between 0 and 100. We then invade the tiling along that are adjacent to the lower edge that has the smallest random number associated with it. By invading this tile, its nearest neighbors (on either side or above it) have become liable for invasion. Hence, we now identify the tile with the smallest random number associated with it among those that may be invaded. After this tile is identified, we invade it. This again changes the set of tiles that may be invaded by including the nearest neighbors of this tile. We repeat this procedure again and again. At some point in the process, the tiling may look like the panel to the left in Figure 4.22. The tiles that may be invaded at the next iteration have random numbers 14, 54, 36, 11, 29, 63, and 44 associated with them. 11 is the smallest number in this list, and it is then invaded as shown in the panel to the right in the figure. This is the invasion percolation model [37], originally constructed for modeling slow drainage in porous media.

In short, in the localized phase when only one hole is growing – see the right panel in Figure 4.21 – the local-load-sharing fiber bundle model in two dimensions is equivalent to the invasion percolation model.

4.2.2
Similarity with the Equal-Load-Sharing Fiber Bundle Model

When the threshold distribution in the local-load-sharing model leads to localization, a singly connected cluster of failed fibers is produced. As we are in two dimensions, islands of surviving fibers will be generated. This is clearly visible in the right panel in Figure 4.21. This is an effect that is entirely missing in the localized failure mode seen in one dimension.

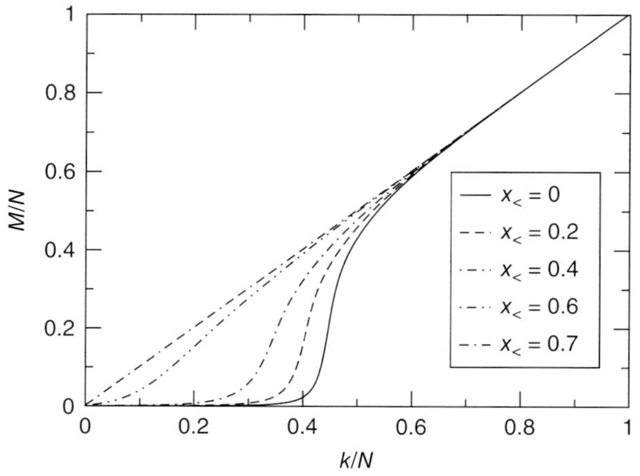

Figure 4.23 The size of the largest hole M in the two-dimensional Local-load-sharing model as a function of the number of broken fibers, k. The threshold distribution was $P(x) = 1 - \exp\,[x_< - x]$ where $x_< \leq x < \infty$ and the number of fibers $N = 256$. Each data set is based on 5000 samples.

When the threshold distribution is *not* leading to localization, as shown in the left panel in Figure 4.21, we also end up with insulated islands of surviving fibers. When the disjoint holes grow large enough, they "bump" into each other and coalesce. This is the percolation transition [38]. Hence, also in this case, insulated islands of surviving fibers will exist in a singly connected hole of failed fibers – just as in the localized case.

This effect is illustrated in Figure 4.23. We show here the size of the largest hole of failed fibers M as a function of the number of broken fibers, k, for different values of the lower cutoff in the threshold distribution $x_<$. For the largest value of $x_< = 0.7$, M grows essentially linearly with k. For $x_< = 0$, we see that M stays close to zero up to a value $k/N \approx 0.4$ when it shoots up. After this, M grows linearly with k. This illustrates well the percolation transition merging the smaller disjoint holes into one large growing hole. In Figure 4.24, we show the percolation transition in M/N versus k/N for different N for the uniform threshold distribution on the unit interval. As N increases, the transition becomes sharper, signaling that this is indeed a phase transition.

In the one-dimensional case, the local-load-sharing rule was described as follows: "the load of an initially missing or failed element is redistributed in equal portions onto its two nearest surviving neighbors" [31]. We have generalized this to two dimensions by assigning the rule that all surviving fibers that are nearest and next-nearest neighbors to a hole of size m share equally the force that these m failed fibers would have carried. Hence, these small islands (see Figure 4.25) – typically single fibers with all their neighbors gone – will then carry the same force since they are all adjacent to the same hole. As a result, *effectively the local-load-sharing fiber bundle model will behave as the equal-load-sharing model!* Due to

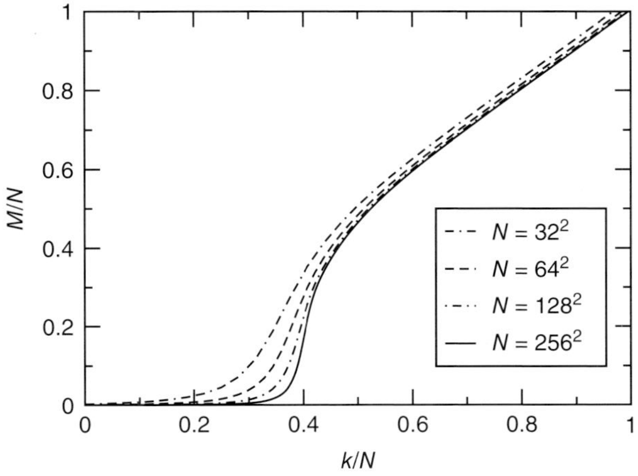

Figure 4.24 The size of the largest hole M in the two-dimensional local-load-sharing model as a function of the number of broken fibers, k for different values of N. The threshold distribution was uniform on the unit interval. Each curve is based on 5000 samples.

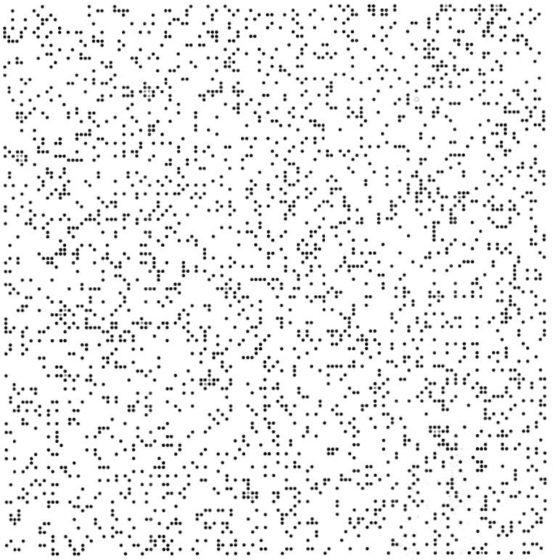

Figure 4.25 This is further on in the breakdown process shown in the right panel in Figure 4.21. This breakdown process has been localized – generating a single hole – from the very start. In this figure, 13 568 fibers have failed and those that remain form isolated islands surrounded by the same "sea" of failed fibers. Hence, the remaining fibers all carry the same stress.

the percolation transition, both the localized and the nonlocalized growth modes of the model will show this closeness.

We show in Figure 4.26 σ versus k/N for the two-dimensional local-load-sharing model for a uniform threshold distribution. We show data for different N ranging from 32^2 to 256^2. The data fall on top of each other. This signals that the

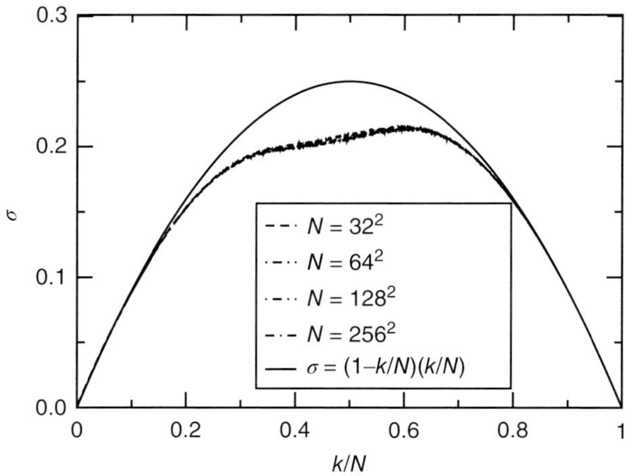

Figure 4.26 σ versus k/N for the two-dimensional local-load-sharing model. The threshold distribution is uniform on the unit interval. The fully drawn graph shows the equal-load-sharing result $\sigma = (1 - k/N)(k/N)$. The figure is based on 5000 samples of each size.

strength of the fiber bundle σ_c converges rapidly to a finite value as $N \to \infty$.[3] We furthermore show in the same figure the behavior of the equal-load-sharing fiber bundle model, which in this case follows the parabola $\sigma = [1 - (k/N)](k/N)$. As we see, the local-load-sharing and equal-load-sharing models are surprisingly close. At the beginning of the failure process, when $k/N < 0.2$, the two models evolve in the same way since the fibers fail because they are weak, not highly stressed since up to this point, there are no large holes. It is much more surprising that there is considerable similarity between the local-load-sharing and equal-load-sharing models when $k/N > 0.8$. This is, however, due to the island effect that we have just discussed: single fibers on the edge of the same percolating hole will carry the same force, and hence, they will fail in ascending order of their thresholds.

The second surprise in Figure 4.26 is that there is data collapse with respect to N. This means that the critical σ, converges to a value *larger than zero* in the limit $N \to \infty$. This behavior is radically different from that found in the one-dimensional local-load-sharing model, $\sigma_c \sim \ln[\ln[N^2]/\ln[N^2] \to 0$ as $N \to \infty$, see Figure 4.13.

We show in Figure 4.27 σ versus k/N for the cumulative threshold distribution $P(x) = 1 - \exp\,[x_< - x]$. We show in the same figure the corresponding curves for the equal-load-sharing model. For $x_< = 0$, there is strong overlap for both the $k/N < 0.2$ and $k/N > 0.8$ regime, whereas the overlap is only good in the $k/N > 0.8$ regime when $x_< = 1$. The strong difference between the local-load-sharing and equal-load-sharing models for $k/N < 0.2$ in this case is due to localization. The local-load-sharing model localizes and behaves as the invasion percolation model, whereas in the equal-load-sharing model there is no localization.

3) This is in contrast to the findings of Domański *et al.* [15], who found $\sigma_c \sim [\ln(N)]^{-0.414}$ as $N \to \infty$.

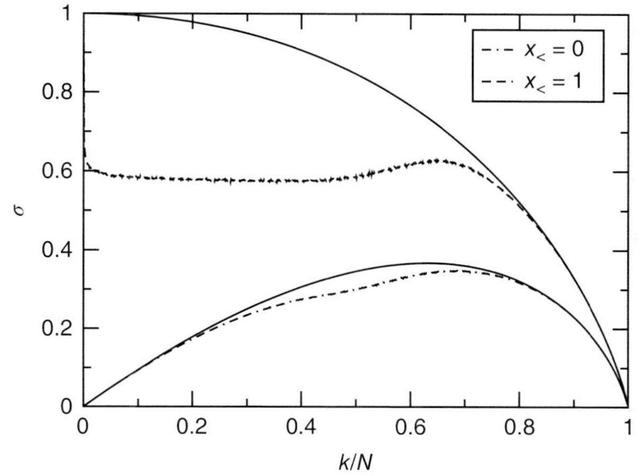

Figure 4.27 σ versus k/N for the two-dimensional local-load-sharing model. The cumulative threshold probability was $P(x) = 1 - \exp\,[x_< - x]$ for $x_< \leq x < \infty$. We also show the equal-load-sharing result $\sigma = (1 - k/N)[x_< - \ln(1 - k/N)]$. The figure is based on 5000 samples for each curve.

It is curious to note that there is a region where the derivative $\mathrm{d}x/\mathrm{d}k$ is positive (roughly $0.5 < k/N < 0.65$) for $x_< = 1$ in the local-load-sharing case. The failure process is stable in this region. In the equal-load-sharing case, there is no such stable region. Hence, stress enhancement at the edges of holes leads to more stability than the lack of stress enhancement. Very curious indeed!

Figure 4.27 should be compared to Figure 4.16 for the one-dimensional local-load-sharing model. We note the dramatic difference between the two.

4.2.3
Burst Distribution

We show the burst distribution for the two-dimensional local-load-sharing model for cumulative threshold distribution $P(x) = 1 - \exp\,[x_< - x]$ with $x_< = 0$ and $x_< = 1$ in Figure 4.28. In both cases, we find that the burst distribution follows the power law

$$D(\Delta) \sim \Delta^{-2.5}, \tag{4.88}$$

that is, the same behavior as found in the equal-load-sharing model, see Equation (3.69).

4.2.4
Upper Critical Dimension

For readers with a background in statistical mechanics, the following speculation may be in place: there could be a critical dimensionality in the local-load-sharing

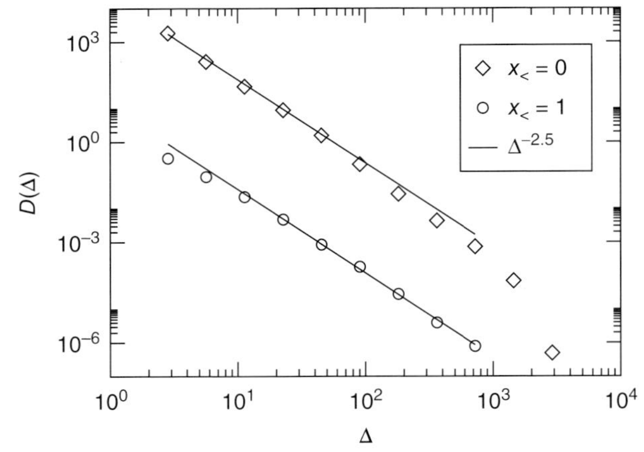

Figure 4.28 Histogram of bursts in the two-dimensional local-load-sharing model. The threshold distribution was $P(x) = 1 - \exp\,[x_< - x]$ where $x_< \leq x < \infty$ and the number of fibers $N = 256$. Each data set is based on 5000 samples.

model where the local-load-sharing and equal-load-sharing models become equal. The results we have presented so far in this section hints at such a scenario: the two-dimensional local-load-sharing model is much closer to the equal-load-sharing model than the one-dimensional local-load-sharing model.

Figure 4.29 shows the σ versus k characteristics of the *three-dimensional* and *four-dimensional* local-load-sharing fiber bundle model [36]. By comparing to Figure 4.26, we see that the higher the dimensionality, the smaller the difference between the local-load-sharing model and the equal-load-sharing model. In Figure 4.30 we show $\Delta\sigma_{area}$ which is the area (integral) of the difference between σ_{LLS} and σ_{ELS} as a function of dimensionality D for $D = 1$, 2, 3, and 5,[4)] for the uniform threshold distribution on the unit interval. The data points fit very well the functional form $\Delta\sigma_{area} \sim D^{-3.4}$. Hence, this figure suggests that the upper critical dimension is infinity. Nevertheless, the no difference between the local-load-sharing fiber bundle and the equal load sharing fiber bundle is very small for $D > 2$.

The local-load-sharing model is extreme in that only the nearest neighbors of the failing fibers are affected. As we will discuss later on in this chapter, a more distributed effect is much more realistic; see Section 4.5. If the extreme local-load-sharing model is very close to the equal-load-sharing model in higher dimensions, it would not be far-fetched that other less extreme model would be so. In three dimensions, fracture models that are far from the fiber bundle models may be constructed. An example is the fuse model, which we will meet later on in this book. Even though they are seemingly very different from the fiber bundle models, in three dimensions they are less extreme than the local-load-sharing model –

4) The five-dimensional data point is based on 40 000 samples of size 6^5.

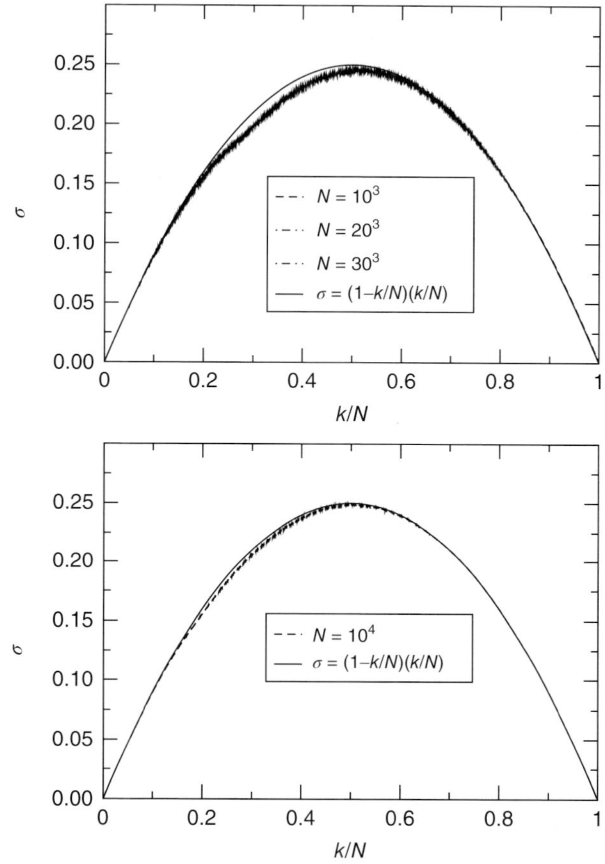

Figure 4.29 Comparing the three-dimensional local-load-sharing model with $N = 10^3, N = 20^3$ and $N = 22^3$ (upper panel) and the four-dimensional local-load-sharing model with $N = 10^4$ (lower panel) to the equal-load-sharing model containing the same number of fibers. The threshold distribution was uniform on the unit interval. The three-dimensional data set have been averaged over 80 000 samples, and the four dimensional data set has been averaged over 30 000 samples (From [36]). This figure should be compared to Figure 4.26.

and as a result, they behave very similarly to the equal-load-sharing model. This is e.g. seen in the burst distribution in the three-dimensional fuse model. Zapperi *et al.* [39] report a burst exponent $\tau = 2.55$, which is very close to the equal-load-sharing value of 5/2.

4.3 The Soft Membrane Model

Is it possible to put some more physics into the local-load-sharing model? As the reader recalls, we described in Section 4.1 how the local-load-sharing model was

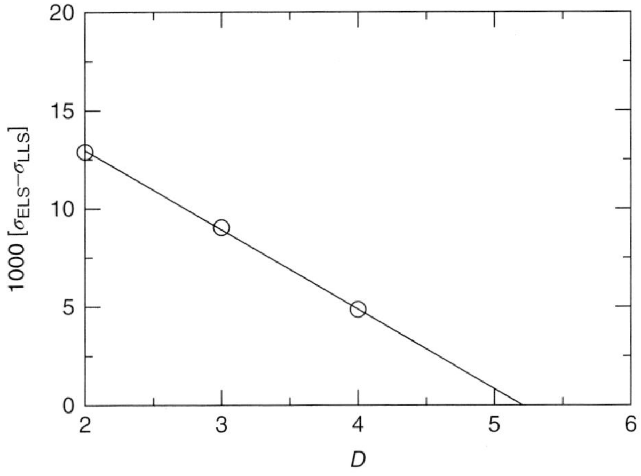

Figure 4.30 $\Delta\sigma_{area}$ as a function of dimensionality D. The straight line is $\Delta\sigma_{area} \sim D^{-3.4}$. The data are based on those presented in Figures 4.26 and 4.29 (Data from [36]).

introduced as a simplified model for the response of a soft clamp to the forces from the fibers. In this Section, we review a recent attempt by Patinet *et al.* [40] to construct an elastic clamp that behaves precisely as the local-load-sharing model in one dimension. However, in two dimensions, it differs from the simple model we have discussed in the previous section.

Patinet *et al.* [40] assume one of the clamps to be a soft membrane, whereas the other is stiff. The forces are then distributed among the fibers by the elastic response of the membrane.

A membrane in two dimensions or a string in one dimension may have two types of elastic responses. One is *bending* elasticity. The other is stretching. We place the membrane along an $x-y$ plane. Deviations in the z direction are denoted $u(x, y)$. If a pressure $p(x, y)$ is applied, the membrane will obey the Kirchhoff–Love equation

$$C\nabla^4 u(x, y) = -p(x, y), \tag{4.89}$$

where C is proportional to the elastic constant E, in the case when the membrane exhibiting bending stiffness only. If, on the other hand, the membrane responds through stretching only without any bending resistance, it obeys the Laplace equation

$$C'\nabla^2 u(x, y) = -p(x, y), \tag{4.90}$$

where C' is proportional to the elastic constant E. We assume in the following that the membrane responds through stretching without any resistance toward bending.

We show in Figure 4.31 an illustration of the soft membrane model. We envision a container in which the pressure may be lowered by moving the piston upward. At the bottom of the container, there is a soft membrane. This membrane is connected to a stiff lower clamp through fibers as shown in Figure 4.31. Each

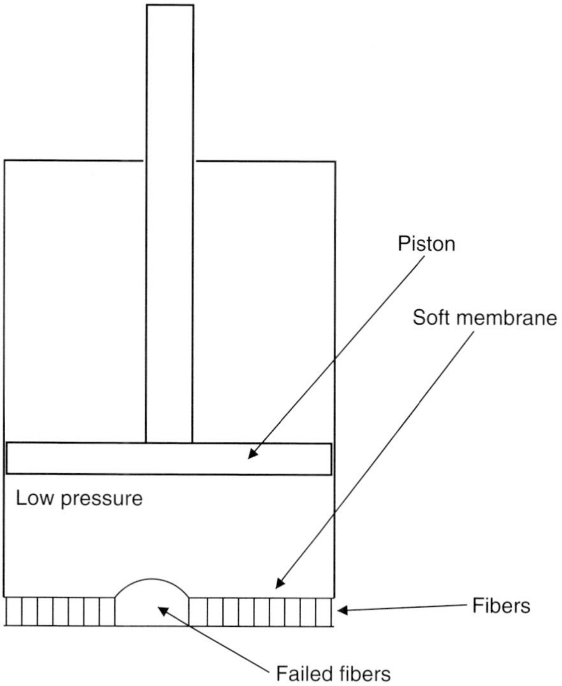

Figure 4.31 Apparatus illustrating the soft membrane model.

fiber has a breaking threshold as described earlier. The fiber bundle is loaded by lowering the pressure in the container – and fibers start failing.

We now assume that the model is one dimensional. Hence, the membrane becomes an elastic string. As shown in Figure 4.32, we assume that there is a hole of size Δx where the fibers have failed. In one dimension, the membrane obeys the equation

$$C' \frac{d^2 u(x)}{dx^2} = p, \tag{4.91}$$

where we assume that the pressure is constant. In the following, we set $C' = 1$. Let us assume that the x axis is placed so that the left end of the hole is at $x = 0$. The boundary conditions are then $u(0) = u(\Delta x) = 0$. We solve Eq. (4.91) to find

$$u(x) = \frac{p}{2} x(\Delta x - x). \tag{4.92}$$

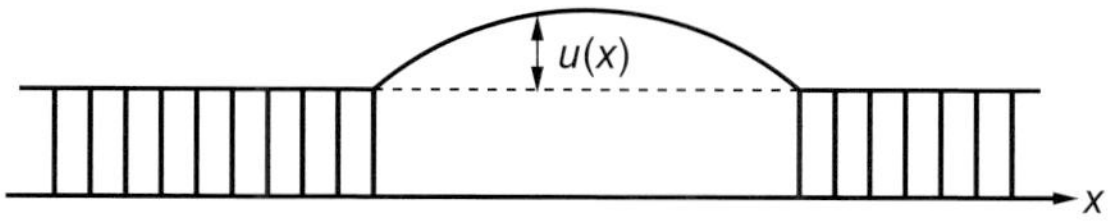

Figure 4.32 Close-up of the soft membrane model where we define $u(x)$.

The force on the fibers at the edges of the hole ($x = 0$ and $x = \Delta x$) is given by

$$\sigma(0) = -\frac{\mathrm{d}u(x)}{\mathrm{d}x}|_{x=0} = \frac{p\Delta x}{2}, \tag{4.93}$$

and

$$\sigma(\Delta x) = \frac{\mathrm{d}u(x)}{\mathrm{d}x}|_{x=\Delta x} = \frac{p\Delta x}{2}. \tag{4.94}$$

We now assume that the fibers are so stiff that we may ignore their elongation in comparison to the deformation of the stretchable membrane. The stress on the edge fibers will then be given by

$$\sigma(0) = \sigma(\Delta x) = p\left[1 + \frac{\Delta x}{2}\right]. \tag{4.95}$$

Fibers that are not at the edge of any hole will experience a stress

$$\sigma = p. \tag{4.96}$$

The pressure p may be converted into the total force on the membrane. We assume that there are N fibers and the distance between them is a. Then $F = Nap$. By rescaling $\sigma \to a\sigma$, Eqs. (4.95) and (4.96) may then be summarized as

$$\sigma = \frac{F}{N}\left[1 + \frac{n_\mathrm{l}}{2} + \frac{n_\mathrm{r}}{2}\right], \tag{4.97}$$

where $n_l = \Delta x_l/a$ and $n_r = \Delta x_r/a$, that is, the number of failed fibers to the left and right of the fiber we are focusing on. This equation should be compared with Eq. (4.6), which defines the local-load-sharing model in one dimension.

This model is straightforward to generalize to two dimensions. We then base the elastic response of the membrane on Eq.(4.90). The force on the fibers at the edge of a hole is then given by

$$\sigma(x, y) = -\vec{n}\cdot\nabla u(x, y), \tag{4.98}$$

where (x, y) lies at the edge of a hole and $\vec{n}$ is the in-plane normal vector at the surface of the hole pointing away from it.

This model is equivalent to the local-load-sharing model in one dimension, but differs from it in two dimensions. We refer the reader to the original article, [40], for further details.

If the reader would like to see if the membrane has bending elasticity rather than stretching elasticity, it can be found in Delaplace *et al.* [41].

4.4
Intermediate-Load-Sharing Models

Local load sharing is quite drastic. Surely fibers further away than just the nearest neighbors of those who fail will feel their plight. Accordingly, one may invent models that distribute the forces carried by the failing fibers further away. We sketch two models here that attempt to "soften" the local-load-sharing rule.

4.4.1 The γ-Model

Hidalgo *et al.* [42] proposed a model where the force that is carried by a broken fiber is distributed to the surviving fibers, according to a power law in the distance from the broken fiber. That is, if the distance between fibers i and j is r_{ij} and fiber i fails, then the additional force on fiber j will be

$$\sigma_{\mathrm{add}} \propto \frac{1}{r_{ij}^{\gamma}}, \tag{4.99}$$

where γ is treated as a parameter. The fibers are placed at the nodes of a square lattice. They introduce a distance cutoff in the range of the force distribution $R_{\max}$. Hence, for those fibers j that have $r_{ij} > R_{\max}$, $\sigma_{\mathrm{add}} = 0$.

Hence, the force added to a surviving fiber j when fiber i fails is then

$$\sigma_{\mathrm{add}} = \sigma_i \begin{cases} 0, & \text{for } r_{ij} > R_{\max}, \\ \frac{Z}{r_{ij}^{\gamma}}, & \text{for } r_{ij} < R_{\max} \end{cases}, \tag{4.100}$$

where

$$Z = \frac{1}{\sum_{i,l}' r_{il}^{-\gamma}}, \tag{4.101}$$

where the sum $\sum'$ only runs over those intact fibers where $r_{il} < R_{\max}$.

Each fiber is given a breaking threshold and the breaking criterion Eq. (4.12) is then used. Hidalgo *et al.* [42] distributed the thresholds according to the Weibull cumulative probability $P(x) = 1 - \exp\,[-x^m]$, where $0 \le x < \infty$ using different values for the Weibull index m.

There is a problem, however, with this model: the force distribution is *not* history independent. History dependence and independence was discussed in 4.1.1. We pointed out that a model should be constructed in such a way that it is possible to determine the force distribution from the distribution of the surviving fibers without knowing the sequence in which the failed fibers failed. The present model breaks this constraint.

This comment aside, the γ-model has a number of interesting properties. We show in Figure 4.33 the strength – critical stress – of the model as a function of the number of fibers it contains, N. Based on this figure, Hidalgo *et al.* [42] proposed that the strength of the model behaves as

$$\sigma_{\mathrm{c}} \sim \frac{1}{\ln(N)}, \tag{4.102}$$

for $\gamma > \gamma_{\mathrm{c}}$ where $\gamma_{\mathrm{c}} \approx 2$. For $\gamma \le \gamma_{\mathrm{c}}$, σ_{c} approaches a constant when $N \to \infty$.

The strength of the two-dimensional local-load-sharing model approaches a constant when $N \to \infty$. This is the same behavior as seen in the γ-model for $\gamma < \gamma_{\mathrm{c}} \approx 2$ and not the behavior seen in this model for $\gamma > \gamma_{\mathrm{c}}$, corresponding to a short range of force redistribution. This is a surprising result.

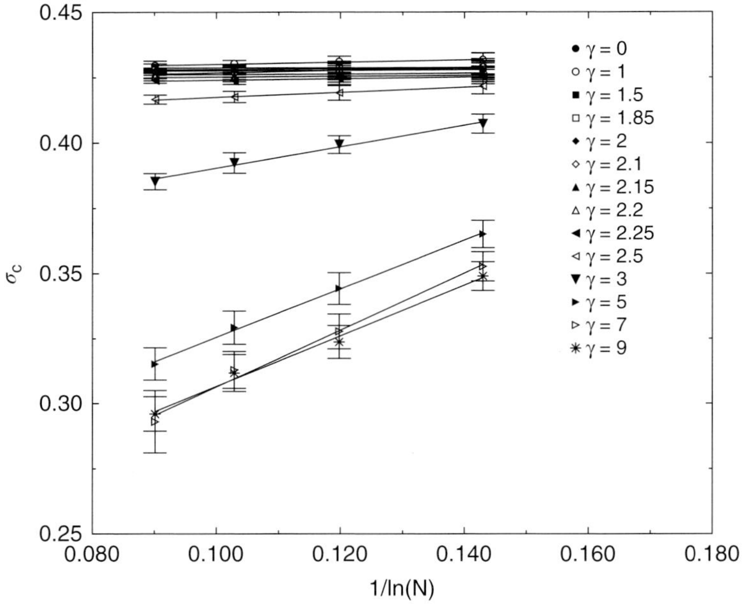

Figure 4.33 Critical stress as a function of the number of fibers N in the γ-model. (From Ref. [42].)

4.4.2
The Mixed-Mode Model

Pradhan *et al.* [43] proposed a model where the equal-load-sharing and local-load-sharing schemes are mixed. That is, when a fiber fails, a fraction g of the load it carried is distributed according to the local-load-sharing rule – that is, to the fibers at the edges of the hole it is part of – and a fraction $1 - g$ of the load is distributed to all the surviving fibers according to the equal-load-sharing rule. Hence, for $g = 0$, the model is the equal-load-sharing model, whereas when $g = 1$, it is the local-load-sharing model.

The model has only been studied in one dimension. Pradhan *et al.* [43] considered the moments of the hole size distribution at the point of failure

$$M_K = \sum_{m=1}^{k_c} m^K p_h(m, k_c). \tag{4.103}$$

By plotting the ratio M_2/M_1 as a function of g, a pronounced peak appeared at a value $g_c = 0.79 \pm 0.01$. At this value, the model crosses over from equal-load-sharing behavior to local-load-sharing behavior.

4.5 Elastic Medium Anchoring

The equal-load-sharing model and the local-load-sharing model are both extremes, but in opposite directions. The equal-load-sharing model, where the clamps are infinitely stiff, does not care about the position of the fibers. The local-load-sharing model, on the other hand, depends strongly on the way the fibers have been placed, along a line, that is, one dimensionally or in the two-dimensional pattern. As was remarked in Section 4.2.4, the higher-dimensional local-load-sharing model is much closer to the equal-load-sharing model than to the one-dimensional local-load-sharing model.

We will in this section introduce a version of the fiber bundle model where forces are transmitted in a physical way through the clamps, which respond elastically [44, 45]. Those with a background in fracture mechanics will see that this model shares some DNA with the Dugdale–Barenblatt cohesive-zone model [46, 47] – the cohesive zone being the region in front of an advancing crack tip. We will use the term *soft clamp fiber bundle model* for the model we present in this section.

In Figure 4.2, we illustrate the model. One clamp is infinitely stiff. The other is soft. That is, it is characterized by a finite Young modulus E and a Poisson ratio ν. Between the two clamps, fibers are placed in a square pattern.[5] Figure 4.2 should be regarded as a one-dimensional cut through the two-dimensional model.[6]

Suppose we now displace the *stiff* clamp by a distance x from the position at which no fiber carries any force. Fiber i then carries a force

$$\sigma_i = -\kappa(u_i - x), \tag{4.104}$$

where u_i is the deformation of the soft clamp at the position of fiber i. We have reintroduced the elastic constant of the fibers, κ. This was set to one early in the chapter. However, in what follows, we need to keep it as a parameter.

We relate the deformation at i to the forces on all the fibers by the Green function for the response of an infinite elastic half space [48-50]

$$u_i = \sum_j G_{ij}\sigma_j, \tag{4.105}$$

where

$$G_{ij} = \frac{1-\nu^2}{\pi E a^2}\int_{-a/2}^{a/2} \frac{dx'dy'}{|\vec{r}_i(x,y) - \vec{r}_j(x',y')|}. \tag{4.106}$$

5) Having one soft and one infinitely stiff clamp is equivalent to having two equally soft clamps. To see this, note that the symmetry plane halfway between the two soft clamps will not change when the fibers are loaded and fail. Hence, one may simply cut the model in two and introduce an infinitely stiff clamp at the position of the symmetry plane.

6) We could, of course, have discussed a one-dimensional version of the model. However, the realistic case is the two-dimensional one. Since we have already seen that the one-dimensional local-load-sharing model is quite different from the two-dimensional one, we only discuss the latter dimensionality for the present model.

We assume that the fibers are placed at the nodes of a square lattice with lattice constant a. Hence, around each fiber i with coordinate $\vec{r}_i = (x_i, y_i)$, there is a Voronoi cell defined by $x - x_i \in [-a/2, +a/2]$ and $y - y_i \in [-a/2, +a/2]$. We use the notation $\vec{r}_i(x, y)$ for the point (x, y) in the Voronoi cell. Hence, $r_{ij} = |\vec{r}_i - \vec{r}_j|$ is the distance between fibers i and j at positions $\vec{r}_i$ and $\vec{r}_j$, respectively, and $|\vec{r}_i(x, y) - \vec{r}_j(x', y')|$ is the distance between any pair of points in the two Voronoi cells surrounding fibers i and j. The integral averages over the area a^2 allocated to each fiber.

The integral (4.106) has been calculated by Love [48] (see Johnson [49]),

$$
\begin{aligned}
\frac{\pi E a^2}{1-\nu^2} G_{ij} = {} & (x_j + a)\ln\left[\frac{(y_j + a) + \sqrt{(y_j + a)^2 + (x_j + a)^2}}{(y_j - a) + \sqrt{(y_j - a)^2 + (x_j + a)^2}}\right] \\
& + (y_j + a)\ln\left[\frac{(x_j + a) + \sqrt{(y_j + a)^2 + (x_j + a)^2}}{(x_j - a) + \sqrt{(y_j + a)^2 + (x_j - a)^2}}\right] \\
& + (x_j - a)\ln\left[\frac{(y_j - a) + \sqrt{(y_j - a)^2 + (x_j - a)^2}}{(y_j + a) + \sqrt{(y_j + a)^2 + (x_j - a)^2}}\right] \\
& + (y_j - a)\ln\left[\frac{(x_j - a) + \sqrt{(y_j - a)^2 + (x_j - a)^2}}{(x_j + a) + \sqrt{(y_j - a)^2 + (x_j + a)^2}}\right].
\end{aligned}
\tag{4.107}
$$

The Green function (4.107) has been derived for an *infinite* half space. However, we assume that the model contains N fibers, and, hence, it is finite. We may, however, still use the infinite half-space Green function in a finite system if we use periodic boundary conditions in both the x and the y directions. We now focus on a given fiber. This fiber will feel the other fibers in the system through the Green function. In addition, it will feel the *copies* of all the fibers – including itself – through the periodicity of the boundary conditions. Hence, in principle, the influence of each fiber will be repeated an infinite number of times, each time at a increasing distance from the fiber we focus on.

In practice, we do not sum up this infinite series of reflections in order to calculate the influence of the fibers on each other. We limit ourselves to one reflection.

From a computational point of view, the soft clamp model is challenging. Numerical techniques have, therefore, been developed to deal with this. One such method that works well is *Fourier acceleration*. We describe this technique in Appendix C.2.

4.5.1
Size Equals Stiffness

The strange title of this subsection hides a very important observation concerning the relation between the stiffness of the soft clamp and the length scales in the model.

We will regard the fiber bundle model as a discretization of a continuum problem. For example, think of two surfaces that are glued together. The glue between the two surfaces, which are our two clamps, is a continuum. We replace this continuum by a set of fibers. Hence, each fiber must then be given an elastic constant κ, which is proportional to a^2.

The Green function (4.106) is proportional to $(Ea)^{-1}$.

The linear size of the system is aL. Hence, by changing the linear size of the system without changing the discretization a, we change $L \to \lambda L$, but leave (Ea) and κ unchanged.

If, on the other hand, we change the discretization without changing the linear size of the system, we simultaneously set $L \to \lambda L, (Ea) \to \lambda(Ea)$, and $\kappa \to \kappa/\lambda^2$.

By defining the rescaled Young modulus

$$e = \frac{Ea}{L}, \tag{4.108}$$

we observe the following: changing e without changing κ is equivalent to changing L – and hence the linear size of the system – while keeping the elastic properties of the system constant [45]. In other words, "size equals stiffness."

In the limit $e \to \infty$, the model becomes identical to the equal-load-sharing fiber bundle model. When $e \to 0$, the other limit, the model does *not* approach the local-load-sharing fiber bundle model. This is easy to see intuitively: the force on a fiber deep inside a fjord of broken fibers will be less strained than a fiber bordering directly a large hole.

4.5.2
Localization in the Soft Clamp Model

Stormo *et al.* [45] considered a rescaled Young modulus in the range $e_{\text{soft}} \leq e \leq e_{\text{stiff}}$ where $e_{\text{soft}} = 2^{-17}$ and $e_{\text{stiff}} = 2^6$. The Poisson ratio was set to $\nu = 1/4$. The fibers had their breaking thresholds distributed according to a uniform distribution on the unit interval.

We show in Figure 4.34 two stages of the failure process in the soft clamp fiber bundle model. In the left panel, localization has not yet set in. In the right panel, localization has occurred. A single, essentially round hole is growing. Curiously, no fibers except on the surface of the hole are failing once localization has set in. It is as if the rest of the system freezes at the point when localization sets in. The reader should compare the shape of this hole to that of the hole the two-dimensional local-load-sharing model produces; see the right panel in Figure 4.21. In the local-load-sharing model, the singly connected hole that develops during localization has a fractal structure. This is not so in this case. Here, it is more or less circular, but with small fluctuations along the rim. It is the long-range elastic response of the soft clamp that causes this.

4.5.3
Asymptotic Strength

Figure 4.34 shows the critical k_c/N as a function of $1/\log_{10}(N)$. Extrapolating the value of k_c/N to the limit $N \to \infty$, gives $(k_c/N)^\infty = 0.16$, that is, a finite value when $e = e_{soft}$. This is in contrast to the one-dimensional local-load-sharing model where $(k_c/N) \to 0$ when $N \to \infty$, see Eq. (4.79). However, it is in line with the two-dimensional local-load-sharing model; see Section 4.2.2. When $e = e_{stiff}$, $(k_c/N)^\infty = 0.5$ in the limit $N \to \infty$. This is the value expected for the equal-load-sharing model.

The corresponding critical stress is shown in the inset in Figure 4.34. The extrapolated values are $\sigma_c^\infty = 0.17$ for $e = e_{soft}$ and $\sigma_c^\infty = 0.25$ for $e = e_{stiff}$. This latter value is the one expected for the equal-load-sharing model.

Figure 4.35 shows the size of the largest hole, $m_{(h)}$ as a function of k for $e = e_{soft}$ and $e = e_{stiff}$.

When $e = e_{soft}$, from around $k/N \approx 0.25$, the largest hole becomes visible and grows essentially linearly from then on. This signals that localization has set in as shown in Figure 4.36. The jumps seen when k/N approaches one stem from the coalescence of holes. When $e = e_{stiff}$, the behavior is consistent with percolation theory: no largest hole dominates until the holes all have

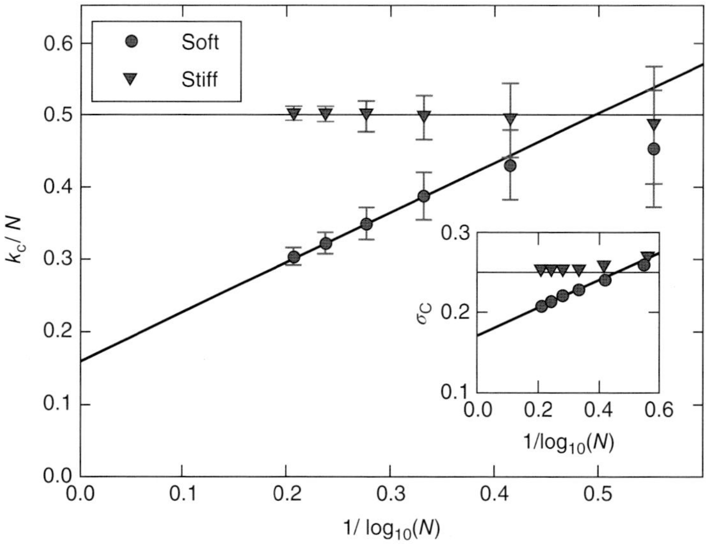

Figure 4.34 Density of the largest cluster of failed fibers, $m_{(h)}$, as a function of k/N for $e = e_{stiff}$ and $e = e_{soft}$. Here $L = 128$. The vertical bar indicates the percolation critical density 0.59274. Here $L = 128$ and 100 samples were generated. (From Ref. [45]).

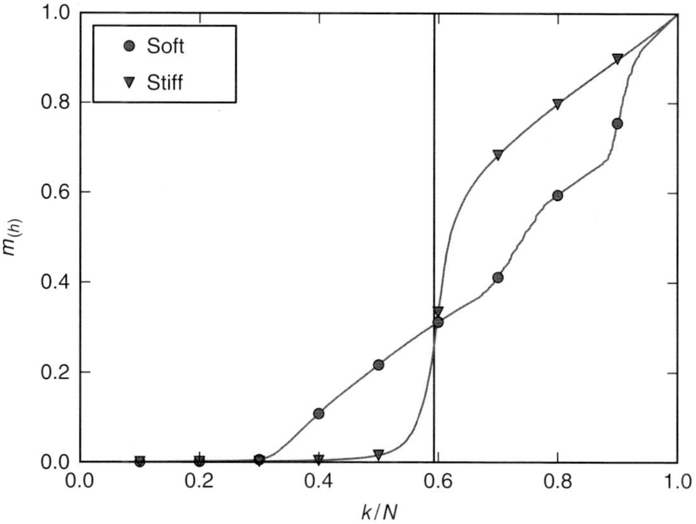

Figure 4.35 The soft clamp model seen from "above." Failed fibers are denoted as black. In the left panel, $k = 4096$ and localization has not yet set in. In the right panel, $k = 8192$ and the only fibers that fail at this point are those on the border of the growing hole. Here, $N = 128^2$ and the rescaled Young modulus was set to e_{soft}. (From Ref. [45].)

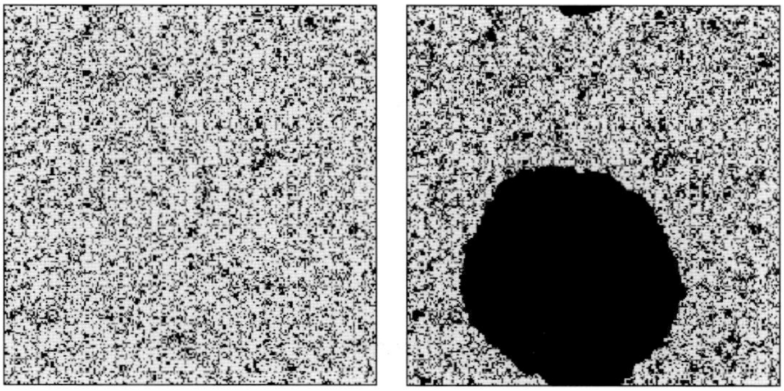

Figure 4.36 Critical k_c/N versus log $_{10}[N]$ for the soft clamp model with the rescaled Young modulus of the clamp being either $e = e_{\text{soft}}$ or $e = e_{\text{stiff}}$. In the case of a stiffer clamp, the value of k_c/N approaches 0.5, which is the equal-load-sharing value. For the soft clamp, critical loading σ_c as a function of $1/\log_{10}[N]$. There is convergence toward a finite value of σ_c as $N \to \infty$. (Adapted from Ref. [51]).

become so large that they "bump" into each other and coalesce. In a completely spatially uncorrelated process, this happens at the percolation critical point, which for site percolation in a square lattice is $k/N = 0.59274\cdots$. This density is marked in the figure, and we see that there is indeed a transition at this value for $e = e_{\text{stiff}}$.

This percolation transition is expected in the equal-load-sharing model. However, as the spatial position of the fibers has no significance in the equal-load-sharing model, the percolation transition plays no role in the system.

4.5.4
Fracture Front Propagation

We end our encounter with the soft clamp model by sketching how it can be used to model fracture front propagation. Let us begin by describing a very elegant experiment pioneered by Schmittbuhl and Måløy [52] on the dynamics and morphology of fracture fronts. They sintered two sandblasted PMMA (traded under the name Plexiglas by DuPont) plates together and then plied them apart slowly from one edge. The sandblasted PMMA plates were originally opaque, but the sintering made them transparent. Where the plates were broken apart again, the opaqueness returned. Hence, it was possible to identify and follow the motion of the fracture front since the plates would be transparent in front of the crack front and opaque behind it. The experimental set is shown in Figure 4.37 and an example of a front is shown in Figure 4.38.

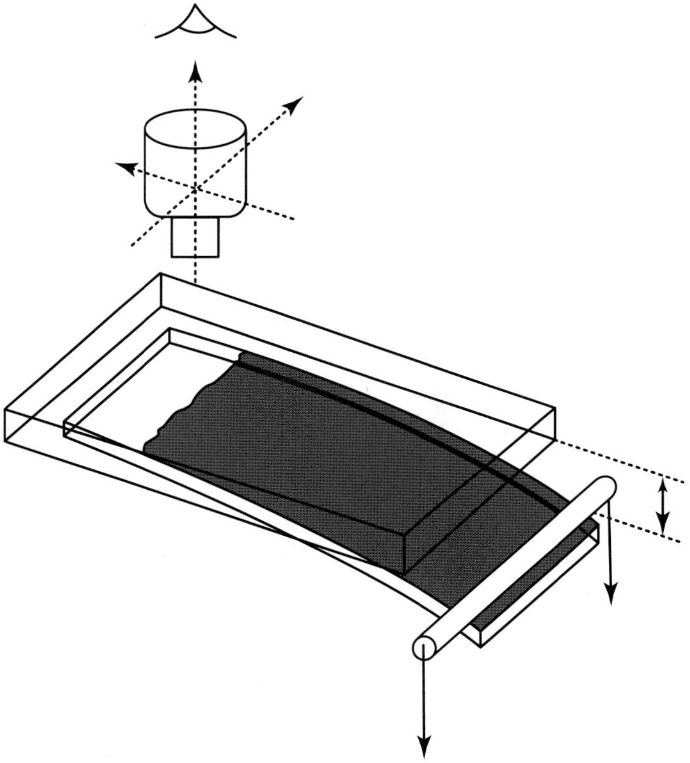

Figure 4.37 The experimental setup by Schmittbuhl and Måløy [52] and later on used by other authors. (Figure credit: K. J. Måløy.)

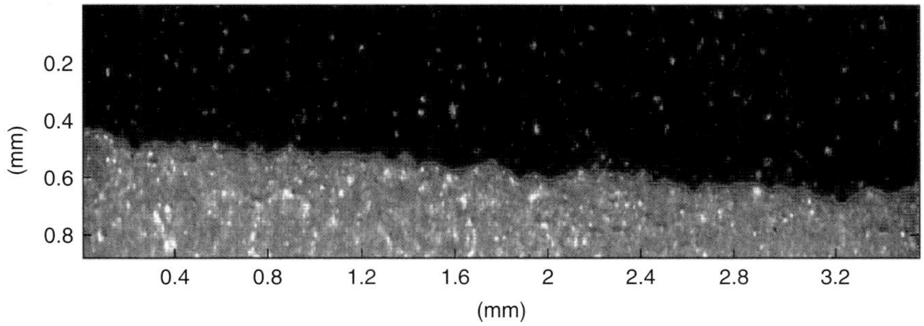

Figure 4.38 The fracture front as seen in the experimental setup shown in Figure 4.37. It moves in the positive *y*-direction. (Photo credit: K. J. Måløy.)

This turned out to be an extremely rich system where an exchange between theoreticians and experimentalists has been very fruitful over a long period, see for example, [53-61].

The soft clamp fiber bundle model has been used to model this system [62-64]. This is done by introducing a gradient in the threshold distribution. The fibers form a square lattice of size $L \times L$. Let the i and j coordinates refer to the (10) and (01) directions, respectively. We then assign a threshold $x_{(i,j)} = gj + r_{(i,j)}$ to the fiber at position (i, j), where $r_{(i,j)}$ is a random number drawn from a uniform distribution on the unit interval. Hence, if we load the system by increasing x, it will fail from one side and a fracture front will develop. This front may then be followed.

We will not go into detail on what has been measured. However, we will just describe one quantity: the velocity distribution [56, 60, 64]. Both the experimental and the fiber bundle data were obtained using the waiting time matrix technique [56]. One defines a matrix where each element corresponds to a fiber (or pixel in the experiment). Initially, each element is set to zero. In the fiber bundle model, we let the number of failed fibers be the "clock" in the model. At each time step, the matrix elements containing the fracture front are incremented by 1. When the front has passed, the matrix contains the time the front spent in each position measured in time steps. We show in Figure 4.39 a graphical rendering of a waiting time matrix by Tallakstad *et al.* [60]. Hence, matrix elements containing low values correspond to rapid front movements and large values correspond to slow, pinned movements. One may then calculate the spatiotemporal map of velocities, $v(i, j)$, by inverting the waiting time recorded in each matrix element. From this map, one then calculates the global average velocity $\langle v \rangle$ and the distribution of local velocities $\omega(v/\langle v \rangle)$. Tallakstad *et al.* [60] found experimentally that this distribution follows a power law,

$$\omega\left(\frac{v}{\langle v \rangle}\right) \sim \left(\frac{v}{\langle v \rangle}\right)^{-\eta}, \quad \text{for} \quad \frac{v}{\langle v \rangle} > 1, \tag{4.109}$$

with the exponent $\eta = 2.55 \pm 0.15$ [60], see the left panel in Figure 4.40.

The result from simulations of the soft clamp model in a threshold gradient is shown in the right panel in Figure 4.40. The measured exponent is $\eta = 2.53$, which is very close to the value obtained in the experiments.

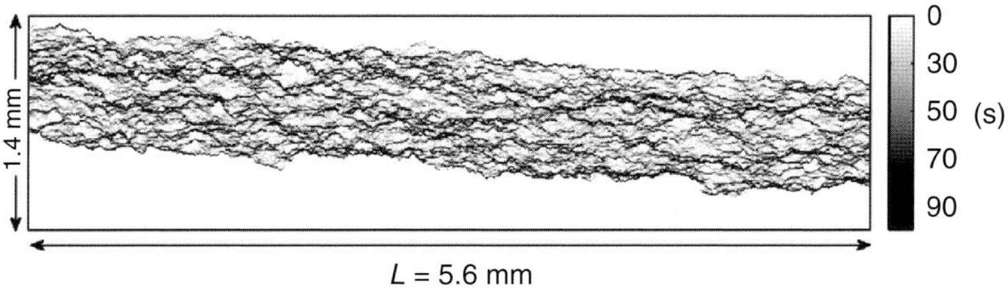

Figure 4.39 A graphical representation of the waiting time matrix for a front moving with an average velocity of 1.36 $\mu m\,s^{-1}$. Each 1/50 s, the position of the front is recorded and added to this figure by adding to the waiting time matrix. Each pixel is gray colored by the time the front has been sitting at that pixel. The bar to the right shows the relation between time and gray shade. By this rendering, the stick-slip – or jerky – motion of the front is clearly visible. (From Ref. [60].)

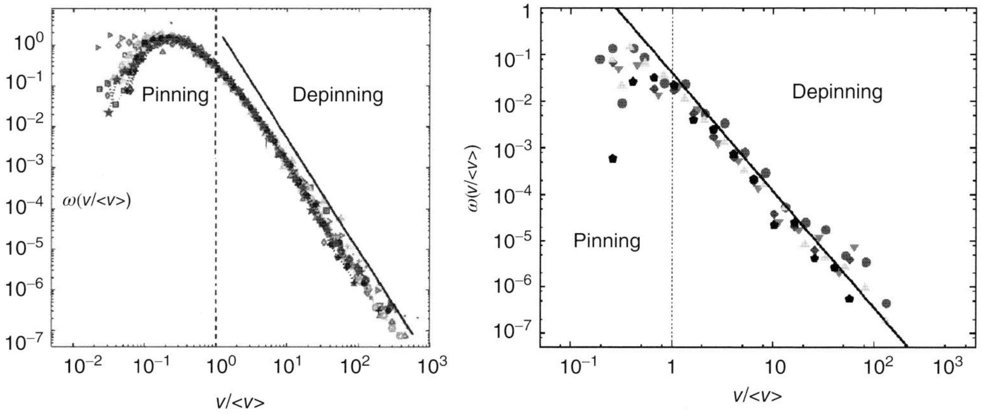

Figure 4.40 Distribution of local velocities scaled by the average velocity from the experimental study by Tallakstad *et al.* [60] (left panel) and from the numerical study by Gjerden *et al.* [64] (right panel). In the upper panel, the normalized velocity distribution based on the waiting time matrix technique is shown. The velocity distribution follows a power law with exponent $\eta = 2.55$ in the depinning regime. The numerical results in the lower panel are based on simulations of sizes $N = 64^2$ and 128^2 with an elastic constant equal to $e = 1.56 \times 10^{-4}$ and 7.81×10^{-4}, respectively. The threshold gradient was in the range $g \in [0.01 - 0.05]$. A fit to the data for $v > \langle v \rangle$ yields a power law described by an exponent $\eta = 2.53$. The "pinning" and "depinning" regimes refer to how the front moves. The pinning regime is characterized by small incremental position changes of the front, and the associated velocities are small, whereas in the depinning regime, it is dominated by the front sweeping over large areas in avalanches. These events are characterized by large velocities.

In fact, the soft clamp fiber bundle model is capable of reproducing essentially all experimental results from the system studied by Måløy and Schmittbuhl and their coworkers at the time when this chapter was written.

5
Recursive Breaking Dynamics

A fiber bundle may be stressed in different ways. When the applied force is increased by infinitesimal increments, the breaking process proceeds such that the weakest fiber among the intact ones always fails. If, however, a *finite* external force is applied, what happens? In the first place, it is clear that all the fibers that cannot withstand the applied load fail. Then, the stress on the surviving fibers increases, which drives further fibers to break and so on.

We model the fiber failure events by this iterative breaking process. It will go on until an equilibrium with some intact fibers is reached or the whole bundle collapses. We now survey the *average* behavior of such breaking processes for a bundle consisting of a large number N of fibers [29, 65–67].

This formulation of the breakdown process in terms of an iterated mapping opens up toward a huge field in mathematics and theoretical physics famous since the 1980s for the concept of *chaos* [68]. The field is called dynamical systems theory (or nonlinear dynamics).

By making a connection between the breakdown process in the fiber bundle and dynamical systems theory, we will be able to understand much better the behavior of the model near collapse. In particular, we will see where singularities develop in the different quantities we monitor and why these singularities are as they are.

In Chapter 3, we noted the similarity between the scaling form of the avalanche size distribution and critical phenomena in statistical physics. This similarity is fortuitous as critical phenomena in statistical physics are the result of spatial correlations. There are no spatial correlation in the equal-load-sharing fiber bundle model. However, through the iterated map formulation of the breakdown process, we will be able to understand why the behavior in the equal-load-sharing model does share so many similarities with critical phenomena.

We extended in Chapter 4 the fiber bundle model in such a way that spatial correlations do occur. In this case, we were dealing with critical phenomena in the full sense of the word. The equal-load-sharing model was recognized as an infinite-dimensional limit of these spatially correlated models. A statistical physicist would use the terminology "mean field limit" for the equal-load-sharing fiber bundle model. The recursive breaking dynamics presented in this chapter exposes the nature of this limit.

The Fiber Bundle Model: Modeling Failure in Materials, First Edition.
Alex Hansen, Per C. Hemmer and Srutarshi Pradhan.

5.1 Recursion and Fixed Points

We assume that an external force F is applied to the fiber bundle, with the applied stress denoted by

$$\sigma = F/N, \tag{5.1}$$

the external load per fiber. We let N_t be the average number of fibers that survive after t steps in the stress redistributing process, with $N_0 = N$. We want to determine how N_t decreases until the degradation process stops. Equal load sharing is assumed.

At a stage during the breaking process when N_t intact fibers remain, the effective stress becomes

$$\sigma_t = N\sigma/N_t. \tag{5.2}$$

With a cumulative probability $P(x)$ of fiber strengths, thus an expected number

$$N\,P(x_t), \quad \text{where } \sigma_t = \kappa x_t, \tag{5.3}$$

of fibers will have thresholds that cannot withstand the load. From now on, we set the elastic constant κ to be equal to 1. In the next step, therefore, the expected number of intact fibers will be

$$N_{t+1} = N - N\,P(N\sigma/N_t). \tag{5.4}$$

In general, the number $N\ P(N\sigma/N_t)$ will not be an integer and should in principle be replaced by its integer part. However, for large N, the deviations from integers will be insignificant in comparison with N_t, except, possibly, in the last stages of the degradation process when merely a few fibers break. By introduction of the relative number of intact fibers through

$$n_t = N_t/N, \tag{5.5}$$

we have

$$x_t = \sigma/n_t, \tag{5.6}$$

and Eq. (5.4) takes the form of a nonlinear recursion relation,

$$n_{t+1} = 1 - P(\sigma/n_t), \tag{5.7}$$

with σ as the control parameter and with $n_0 = 1$ as the start value.

Alternatively, we can set up a recursion for x_t, the effective stress σ/n_t after t iterations:

$$x_{t+1} = \frac{\sigma}{1 - P(x_t)}, \tag{5.8}$$

with $x_0 = \sigma$ as the initial value. By (5.6) the two recursion schemes may be mapped onto each other.

In general, it is not possible to solve nonlinear iterations like (5.7) or (5.8) analytically. A model with a uniform threshold distribution is, however, an exception, as we will see in Sections 5.2.2 and 5.2.3.

In nonlinear dynamics, the character of an iteration is primarily determined by its *fixed points*. A fixed point is a value or a point that is mapped onto itself by the iteration. We are, therefore, interested in possible fixed points n^* of (5.7), which satisfy

$$n^* = 1 - P(\sigma/n^*). \tag{5.9}$$

Correspondingly, fixed points x^* of the iteration (5.8) must satisfy

$$x^* = \frac{\sigma}{1 - P(x^*)}, \tag{5.10}$$

which may be written

$$F = N\sigma = Nx^*(1 - P(x^*)). \tag{5.11}$$

This is precisely the relation (2.12) between stress and strain. Therefore, the equilibrium value of x, for a given external stress σ, is a fixed point. This presupposes that the stress is less than the maximal bundle strength per fiber, $\sigma < \sigma_c$, because only then will the fixed-point Eq. (5.11) have a real solution for x^*. In other words, only then will an equilibrium with some intact fibers exist.

For $\sigma < \sigma_c$, does a start value x_0 always iterate toward the fixed point x^*? A necessary condition for this to be the case is that the fixed point is *attractive*. To check this, one linearizes the iteration (5.8) around the fixed point:

$$x_{t+1} - x^* = \frac{\sigma}{1 - P(x^*)} + \frac{\sigma p(x^*)}{[1 - P(x^*)]^2}(x_t - x^*) - x^* = A(x_t - x^*), \tag{5.12}$$

with

$$A = \frac{\sigma p(x^*)}{[1 - P(x^*)]^2} = \frac{x^* p(x^*)}{1 - P(x^*)}, \tag{5.13}$$

using (5.10). The fixed point is attractive (stable) if $|A| < 1$ and repulsive (unstable) if $|A| > 1$.

In the generic case, the load curve $F(x) = Nx[1 - P(x)]$ has precisely one parabolic maximum (at x_c). The gradient, therefore, satisfies

$$\frac{\mathrm{d}F(x)}{\mathrm{d}x} = N[1 - P(x) - xp(x)] = \begin{cases} > 0 & \text{for } x < x_c \\ < 0 & \text{for } x > x_c. \end{cases} \tag{5.14}$$

For a given applied stress $\sigma = F/N$ less than the maximum strength per fiber σ_c, the fixed-point Eq. (5.10) will have *two* solutions for x^*. It follows from (5.13) and (5.14) that the fixed point $x^* < x_c$ is attractive ($|A| < 1$), while the fixed point $x^* > x_c$ is repulsive. Since $A > 0$, the stable fixed point is approached from one side (through monotonically increasing values x_t). Since $n_t = \sigma/x_t$, Eq. (5.6), the number of intact fibers will approach its fixed point through monotonically *decreasing* values n_t.

The stable fixed points n^* and x^* are related by

$$n^* = \sigma/x^*, \tag{5.15}$$

a consequence of the $t \to \infty$ limit in the relation (5.6).

Let us illustrate these general results by an example. For fiber strengths with the uniform threshold distribution (2.16) with dimensionless distance, that is,

$$P(x) = \begin{cases} x & \text{for } 0 \le x \le 1 \\ 1 & \text{for } x > 1, \end{cases} \tag{5.16}$$

the fixed-point Eqs. (5.9) and (5.10) take the form

$$n^{*2} - n^* + \sigma = 0 \quad \text{and} \quad x^{*2} - x^* + \sigma = 0, \tag{5.17}$$

with solutions

$$n^* = \frac{1}{2} \pm (\sigma_c - \sigma)^{1/2} \tag{5.18}$$

and

$$x^* = \frac{1}{2} \mp (\sigma_c - \sigma)^{1/2}. \tag{5.19}$$

Here $\sigma_c = 1/4$, the critical value of the applied stress beyond which the bundle fails completely. In (5.18) and (5.19), the upper signs give $n^* > n_c$ and $x^* < x_c$, which correspond to stable fixed points. The lower signs correspond to unstable fixed points.

To track down the approach very near a fixed point, we note that close to a stable fixed point the iterated quantity changes by tiny amounts, so that one may expand in the differences $n_t - n^*$ and $x^* - x_t$. For a model with uniform distribution of the thresholds, the recursion relation (5.7),

$$n_{t+1} = 1 - \sigma/n_t, \tag{5.20}$$

gives to linear order

$$n_{t+1} - n^* = \frac{\sigma}{n^*} - \frac{\sigma}{n_t} = \frac{\sigma}{n_t n^*}(n_t - n^*) \simeq \frac{\sigma}{n^{*2}}(n_t - n^*). \tag{5.21}$$

Thus, the fixed point is approached monotonously with exponentially decreasing steps:

$$n_t - n^* \propto e^{-t/\tau}, \tag{5.22}$$

with a relaxation parameter

$$\tau = 1/\ln\,(n^{*2}/\sigma) = 1/\ln\left[\left(\frac{1}{2} + \sqrt{\frac{1}{4} - \sigma}\right)^2 \Big/ \sigma\right]. \tag{5.23}$$

For the critical load, $\sigma = \sigma_c = \frac{1}{4}$, the argument of the logarithm is 1, so that apparently τ is infinite. More precisely, the dominating term in (5.23) for $\sigma \to \sigma_c$ is

$$\tau \simeq \frac{1}{4}\,(\sigma_c - \sigma)^{-\theta} \quad \text{with } \theta = \frac{1}{2}. \tag{5.24}$$

The divergence is a clear indication that the character of the breaking dynamics changes when the bundle goes critical. We now look into the dynamics at or very near criticality.

5.2 Recursive Dynamics Near the Critical Point

The critical load for the bundle with the uniform threshold distribution is, as noted earlier, $\sigma_c = \frac{1}{4}$, corresponding to $x_c = \frac{1}{2}$. Under the critical load, one-half of the fibers have ruptured, $n_c^* = 1/2$, as is clear from the fixed-point expression (5.18) with $\sigma = \sigma_c$, or by using (5.15).

Stable fixed points represent equilibrium values: $n(\sigma) = n^*(\sigma)$. The stable fixed point (5.18) implies that the relative number $n(\sigma)$ of unbroken fibers approaches its critical value $n_c^* = n(\sigma_c)$ as follows:

$$n^*(\sigma) - n^*(\sigma_c) \simeq (\sigma - \sigma_c)^\alpha \quad \text{with } \alpha = \frac{1}{2}. \tag{5.25}$$

The difference $\mathcal{O} = n^*(\sigma) - n^*(\sigma_c)$ acts as a kind of order parameter. A negative value of $\mathcal{O}$ signals complete bundle breakdown, while for positive values the bundle survives with a sufficient number of intact fibers. When the load is increased through its critical value, $\mathcal{O}$ changes from finite positive values to zero at the critical point, and then abruptly to the negative value $\mathcal{O} = -n^*(\sigma_c)$, since now all fibers have failed, $n^* = 0$. (Figure 5.1)

One can also define a breakdown susceptibility χ as the change in $n^*(\sigma)$ by an infinitesimal increment of the applied stress σ:

$$\chi = -\frac{dn^*}{d\sigma}. \tag{5.26}$$

The Eq. (5.18) for the stable fixed point gives immediately

$$\chi = \frac{1}{2}(\sigma_c - \sigma)^{-\beta} \quad \text{with } \beta = \frac{1}{2}. \tag{5.27}$$

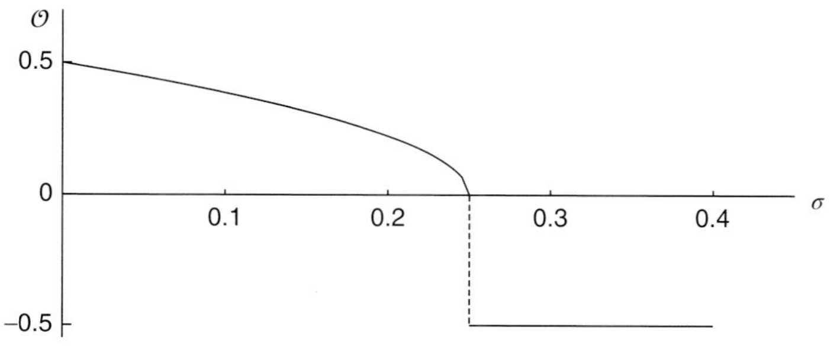

Figure 5.1 The order parameter $\mathcal{O} = n^*(\sigma) - n^*(\sigma_c)$ as function of the stress σ, for the uniform threshold distribution.

Let us reconsider how iterations approach the fixed point n^* when the load is *critical*, $\sigma = 1/4$. In terms of the difference

$$\Delta n_t = n_t - n_c^*, \tag{5.28}$$

the recursion relation (5.20) may be written $\Delta n_{t+1} = \Delta n_t/(1 + 2\Delta n_t)$, or

$$\Delta n_{t+1}^{-1} = \Delta n_t^{-1} + 2. \tag{5.29}$$

Thus, the inverse function Δn_t^{-1} increases by 2 at each iteration. Since the start value is $\Delta n_0^{-1} = (1 - \frac{1}{2})^{-1} = 2$, we obtain $\Delta n_t^{-1} = 2 + 2t$, that is,

$$\Delta n_t = n_t - n_c^* = \frac{1}{2t + 2}. \tag{5.30}$$

Hence, one sees at criticality a *power-law* approach to the fixed point for large t,

$$n_t - n_c^* \propto t^{-\eta}, \quad \text{with } \eta = 1, \tag{5.31}$$

instead of the usual rapid exponential approach (5.22). This *critical slowing down* is a robust characterization of the critical state.

In a similar manner, we can find how x_t approaches its fixed point at criticality. By (5.15), we have $x_c^* - x_t = \sigma_c(1/n_c^* - 1/n_t)$. Insertion of n_t from (5.30) gives

$$x_c^* - x_t = \frac{1}{2t + 4}, \tag{5.32}$$

essentially the same power-law decay for large t, $x_c^* - x_t \propto t^{-1}$, as we saw for $n_t - n_c^*$.

5.2.1
Universality

The critical properties obtained are for the uniform distribution of the breaking thresholds, and the natural question is how general the results are. We start by a spot check on universality through considering a different distribution of fiber strengths.

A *linearly increasing distribution* on the interval (0,1),

$$p(x) = \begin{cases} 2x, & 0 \le x \le 1 \\ 0 & x > 1, \end{cases} \tag{5.33}$$

is shown in Figure 5.2. For simplicity, nondimensional variables are used.

By (2.13) the average total force per fiber (Figure 5.3),

$$F(x)/N = \begin{cases} x(1 - x^2) & 0 \le x \le 1 \\ 0 & x > 1, \end{cases}, \tag{5.34}$$

shows that the critical point is

$$x_c = \frac{1}{\sqrt{3}}, \quad \sigma_c = \frac{2}{3\sqrt{3}}. \tag{5.35}$$

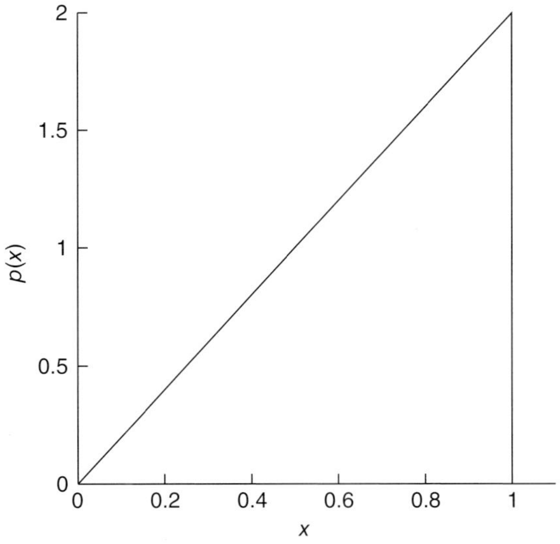

Figure 5.2 The increasing fiber strength distribution (5.33).

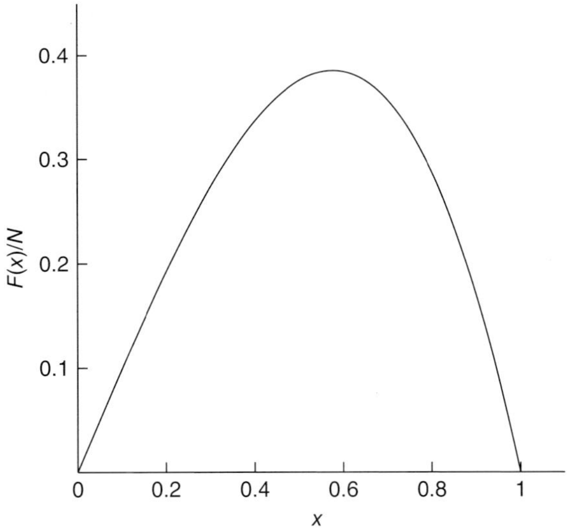

Figure 5.3 The average total force per fiber, $F(x)/N$, for the increasing strength distribution (5.33).

In this case, the recursive dynamics (5.7) and (5.8) take the form

$$n_{t+1} = 1 - \frac{\sigma^2}{n_t^2} \quad \text{and} \quad x_{t+1} = \frac{\sigma}{1 - x_t^2}. \tag{5.36}$$

The fixed-point equation for x^* is $x^* = \sigma/(1 - x^{*2})$, which may be written

$$x^* - x^{*3} = \sigma, \tag{5.37}$$

a cubic equation in x^*. Therefore, there exist three solutions x^* for each value of σ. In Figure 5.4, the relation between σ and x^* is plotted for real values. For $|\sigma| \leq 2/3\sqrt{3}$, the three zeroes of the cubic equation are all real, while for $|\sigma| > 2/3\sqrt{3}$ one zero is real, so that the other two must be complex. In any case, only the positive quadrant can be of physical relevance, with bundle stability for

$$0 \leq \sigma \leq \sigma_c = 2/3\sqrt{3}, \tag{5.38}$$

One of the two zeroes x^* in this region is smaller than x_c, the other one larger.

We linearize around each fixed point to check whether the fixed point is stable or not:

$$\begin{aligned} x^* - x_{t+1} &= \frac{\sigma}{1 - x^{*2}} - \frac{\sigma}{1 - x_t^2} = \frac{\sigma(x^{*2} - x_t^2)}{(1 - x^{*2})(1 - x_t^2)} \\ &\simeq (x^* - x_t)\,\frac{2\sigma x^*}{(1 - x^{*2})^2} = (x^* - x_t)\frac{2x^{*2}}{1 - x^{*2}}, \end{aligned} \tag{5.39}$$

by means of (5.37). This may be written

$$\frac{x^* - x_{t+1}}{x^* - x_t} = 1 + \frac{3x^{*2} - 1}{1 - x^{*2}}. \tag{5.40}$$

For the fixed point larger than $x_c = 1/\sqrt{3}$ the ratio (5.40) is larger than 1; this fixed point is, therefore, unstable. The smaller fixed point, for which $x^* < x_c$, is *stable*.

By (5.36), the fixed points n^* for the relative number of intact fibers satisfy

$$n^{*2} - n^{*3} = \sigma^2. \tag{5.41}$$

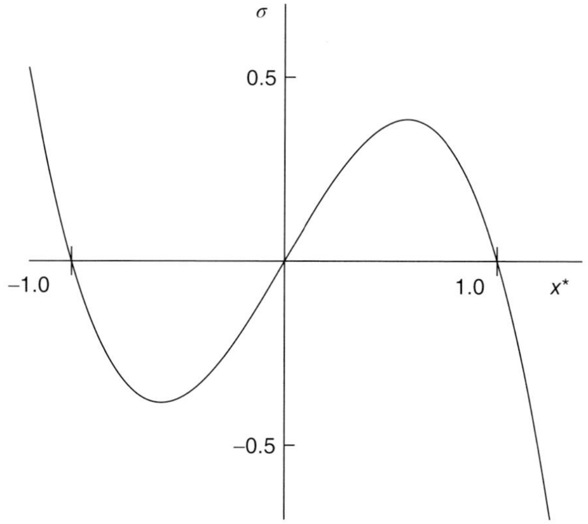

Figure 5.4 The relation between the fixed-point value x^* and the applied stress σ for the increasing strength distribution (5.33). The maximum and minimum values are $\sigma = \pm 2/3\sqrt{3}$ and occur at $x^* = \pm 1/\sqrt{3}$.

For $0 \leq n^*$, the left-hand side of (5.42) is a positive function of n^* with a maximum equal to 4/27, occurring at $n^* = 2/3$. For the critical load, $\sigma_c = 2/3\sqrt{3}$, the only real and positive solution of (5.41) is, therefore,

$$n_c^* = \frac{2}{3}. \tag{5.42}$$

In a similar way as for x^*, one shows that for $\sigma < \sigma_c$ there will be an unstable fixed point with $n^* < n_c^*$ and a stable one with $n^* > n_c^*$.

To find the number of intact fibers in the neighborhood of the critical point, we insert $n = \frac{2}{3} + (n - n_c)$ into (5.41), with the result

$$\begin{aligned} \frac{4}{27} - (n - n_c)^2 - (n - n_c)^3 = \sigma^2 &= (\frac{2}{3\sqrt{3}} + \sigma - \sigma_c)^2 \\ &= \frac{4}{27} + \frac{4}{3\sqrt{3}}(\sigma - \sigma_c) + (\sigma - \sigma_c)^2. \end{aligned} \tag{5.43}$$

To leading order we have

$$(n - n_c)^2 = \frac{4}{3\sqrt{3}}(\sigma_c - \sigma). \tag{5.44}$$

Hence for $\sigma \leq \sigma_c$ the order parameter behaves as

$$n(\sigma) - n_c \propto (\sigma_c - \sigma)^\alpha, \quad \alpha = \frac{1}{2}, \tag{5.45}$$

in accordance with (5.25). The breakdown susceptibility $\chi = -\,\mathrm{d}n/\mathrm{d}\sigma$ will therefore have the same critical behavior,

$$\chi \propto (\sigma_c - \sigma)^{-\beta}, \quad \beta = \frac{1}{2} \tag{5.46}$$

as for the model with the uniform distribution of fiber strengths.

Let us also investigate how the stable fixed point is approached. From (5.36) we find

$$n_{t+1} - n^* = \frac{\sigma^2}{n^{*2}} - \frac{\sigma^2}{n_t^2} = \frac{\sigma^2}{n^{*2}n_t^2}\,(n_t^2 - n^{*2}) \simeq (n_t - n^*)\,\frac{2\sigma^2}{n^{*3}} \tag{5.47}$$

near the fixed point. Hence, the approach is exponential,

$$n_t - n^* \propto e^{-t/\tau} \quad \text{with} \quad \tau = \frac{1}{\ln\,(n^{*3}/2\sigma^2)}. \tag{5.48}$$

At the critical point, where $n_c^* = 2/3$ and $\sigma_c = 2/3\sqrt{3}$, the argument of the logarithm equals 1, so that τ diverges when the critical state is approached. The divergence is easily seen to be of the same form, $\tau \propto (\sigma_c - \sigma)^{-1/2}$, as for the model with a uniform threshold distribution, Eq. (5.24).

To find the correct behavior of the distance to the critical point, $\Delta n_t = n_t - n_c = n_t - 2/3$, *at criticality*, we use the iteration (5.36) with $\sigma = \sigma_c$,

$$n_{t+1} = 1 - \frac{4/27}{n_t^2}, \quad \text{or} \quad \Delta n_{t+1} = \frac{1}{3} - \frac{4/27}{\left(\frac{2}{3} + \Delta n_t\right)^2}. \tag{5.49}$$

Near the fixed point, the deviation $\Delta n_t = n_t - n^*$ is small. An expansion to second order in Δn_t yields

$$\Delta n_{t+1} = \Delta n_t - \frac{9}{4}\Delta n_t^2, \tag{5.50}$$

which is satisfied by

$$\Delta n_t = \frac{4}{9t} + \mathcal{O}(t^{-2}). \tag{5.51}$$

The slow critical relaxation toward the fixed point, $n_t - n_c \propto t^{-1}$ for large t, is the same as for the uniform threshold distribution, formula (5.31).

In conclusion, we have found that the model with a linearly increasing distribution of the fiber strengths possesses the same critical exponent as the model with the uniform distribution. This suggests that the critical properties are universal, which we are now going to prove through general arguments.

5.2.1.1 General Threshold Distribution

In the generic case, the macroscopic force function $F(x)$ has a single parabolic maximum corresponding to the critical point at x_c. In a subcritical situation near criticality, therefore, the force per fiber function $\sigma(x) = x[1 - P(x)]$ has the expansion

$$\sigma(x) = \sigma_c - \frac{1}{2}\sigma''(x_c)(x_c - x)^2 = \sigma_c - \frac{1}{2}[2p(x_c) + x_c p'(x_c)]\,(x_c - x)^2. \tag{5.52}$$

Since $n(\sigma)$ is determined by $x(\sigma)$, relation (5.15), the force function is also a quadratic function of the fraction of intact fibers near criticality,

$$\sigma = \sigma_c - C(n - n_c)^2, \tag{5.53}$$

with a positive C. Thus, the order parameter vanishes as follows:

$$\mathcal{O} = n(\sigma) - n(\sigma_c) \propto (\sigma_c - \sigma)^\alpha, \quad \alpha = \frac{1}{2}, \tag{5.54}$$

when σ approaches σ_c from below. This is in accordance with what we found for the two special distributions of fiber strengths.

Equation (5.54) implies also that the breakdown susceptibility $\chi = -\mathrm{d}n(\sigma)/\mathrm{d}\sigma$ diverges as $(\sigma_c - \sigma)^{-\beta}$ with the exponent $\beta = \frac{1}{2}$, as for the two special distributions.

Finally, we consider how the stable fixed point is approached. The iteration (5.7) yields

$$n_{t+1} - n^* = P\left(\frac{\sigma}{n^*}\right) - P\left(\frac{\sigma}{n_t}\right) = P\left(\frac{\sigma}{n^*}\right) - P\left(\frac{\sigma}{n^* + (n_t - n^*)}\right). \tag{5.55}$$

When n_t is close to the fixed point n^*, we expand the right-hand side of (5.55) to first order in the small quantity $n_t - n^*$, obtaining

$$n_{t+1} - n^* = (n_t - n^*)\, p(\sigma/n^*)\, \sigma/n^{*2} \tag{5.56}$$

to first order. This guarantees an exponential relaxation to the fixed point,

$$n_t - n^* \propto e^{-t/\tau}, \tag{5.57}$$

with

$$\tau = 1/\ln\left(\frac{n^{*2}}{\sigma\, p(\sigma/n^*)}\right) = 1/\ln\left(\frac{\sigma}{x^{*2}p(x^*)}\right), \tag{5.58}$$

using $n^* = \sigma/x^*$, Eq. (5.6). Since the fixed-point values are equilibrium values, the asterisks are superfluous now. By the critical condition (2.15), $\sigma_c = x_c^2 p(x_c)$, the argument of the logarithm in (5.58) is 1 at criticality. Thus, the expression (5.58) for the relaxation parameter yields $\tau = \infty$, which implies that the critical relaxation is *not* exponential. The near-critical behavior of the relaxation parameter τ follows from (5.58) by an expansion to first order in $x_c - x$. Since by (5.52) σ deviates from σ_c to second order in $x_c - x$, we may to first order replace σ by σ_c. Moreover,

$$\begin{aligned}\frac{\sigma_c}{x^2p(x)} &= \frac{\sigma_c}{x_c^2p(x_c) + (x - x_c)[2x_cp(x_c) + x_c^2p'(x_c)]}\\ &= \frac{1}{1 + (x - x_c)\sigma_c^{-1}[2x_cp(x_c) + x_c^2p'(x_c)]}\end{aligned} \tag{5.59}$$

Insertion into (5.58) yields

$$\tau = \frac{\sigma_c}{2x_cp(x_c) + x_c^2p'(x_c)}(x_c - x)^{-1}. \tag{5.60}$$

Finally, use of (5.52) gives

$$\tau = \frac{\sigma_c/x_c}{\sqrt{2x_cp'(x_c) + 4p(x_c)}}\,(\sigma_c - \sigma)^{-1/2}. \tag{5.61}$$

Hence, the square-root divergence when $\sigma \to \sigma_c$ is universal. For the uniform threshold distribution, $p(x) = 1$, the previous result (5.24) is obtained.

To study the relaxation *at criticality*, we must expand (5.55) to second order in $(n_t - n^*)$:

$$n_{t+1} - n^* = n_t - n^* - C(n_t - n^*)^2, \tag{5.62}$$

with a positive constant C. This is satisfied by

$$n_t - n^* = C^{-1}t^{-1} + \mathcal{O}(t^{-2}), \tag{5.63}$$

the same slow relaxation as we have seen for the two special cases.

In conclusion, we have demonstrated universality: all equal-load-sharing models with a fiber strength distribution that corresponds to a parabolic maximum of the load curve $F(x)$, have the same critical behavior.

5.2.2
Postcritical Relaxation

The iterative breaking process considered ends with one of two possible results. Either the whole bundle breaks down or an equilibrium situation with a finite number of intact fibers is reached. The final fate depends on whether the external stress σ on the bundle is postcritical ($\sigma > \sigma_c$), precritical ($\sigma < \sigma_c$), or critical $\sigma = \sigma_c$. We now investigate the total number t_f of iterative steps t necessary to reach the final state and start with the *postcritical* situation. For the model with the uniform threshold distribution (5.16), which we consider first, this means that $\sigma > \sigma_c = 1/4$. Intuitively, it seems clear that the higher the external stress, the fewer are the number of steps necessary to reach complete collapse. We will now make that expectation quantitative.

5.2.2.1 Uniform Threshold Distribution

For the model with the uniform threshold distribution (5.16), we can explicitly and exactly follow the path of iteration. We introduce a measure ε for how much the load σ exceeds the critical load σ_c by

$$\varepsilon = \sigma - \sigma_c = \sigma - \frac{1}{4}, \tag{5.64}$$

where ε is positive. The basic iteration formula (5.7) is

$$n_{t+1} = 1 - \sigma/n_t. \tag{5.65}$$

The fraction n_t of intact fibers will decrease under the iteration, and we see from (5.65) that if n_t reaches the value σ or a smaller value, the next iteration yields $n_{t+1} = 0$ or a negative value, that is, complete bundle breakdown. We wish to find how many iterations, t_f, are needed to reach this stage.

For that purpose, we solve the nonlinear iteration (5.65) by converting it into a linear iteration by means of two transformations. Into (5.65),

$$n_{t+1}n_t = n_t - \sigma = n_t - \frac{1}{4} - \varepsilon, \tag{5.66}$$

we introduce first

$$n_t = \frac{1}{2} - y_t\sqrt{\varepsilon}, \tag{5.67}$$

with the result

$$2\sqrt{\varepsilon} = \frac{y_{t+1} - y_t}{1 + y_{t+1}y_t}. \tag{5.68}$$

As a second transformation, put

$$y_t = \tan v_t, \tag{5.69}$$

with the result

$$2\sqrt{\varepsilon} = \frac{\tan v_{t+1} - \tan v_t}{1 + \tan v_{t+1}\ \tan v_t} = \tan(v_{t+1} - v_t). \tag{5.70}$$

Hence we have obtained the linear iteration

$$v_{t+1} - v_t = \arctan(2\sqrt{\varepsilon}), \tag{5.71}$$

with solution

$$v_t = v_0 + t \arctan(2\sqrt{\varepsilon}). \tag{5.72}$$

The iteration starts with all intact fibers, that is, $n_0 = 1$, which by (5.67) and (5.69) corresponds to $y_0 = -1/2\sqrt{\varepsilon}$ and $v_0 = -\arctan(1/2\sqrt{\varepsilon})$. With the constant in

(5.72) now determined, we can express the solution in terms of the original variable:

$$n_t = \frac{1}{2} - \sqrt{\varepsilon}\tan[-\arctan(1/2\sqrt{\varepsilon}) + t\ \arctan(2\sqrt{\varepsilon})]. \tag{5.73}$$

We saw that when n_t obtains a value in the interval $(0, \sigma)$, the next iteration gives $n_{t+1} \leq 0$, complete bundle failure. Using (5.73), we can determine how many iterations are needed to reach the values 0 and σ, and to these values we add 1 for the final iteration. Reaching the smallest value 0 gives an *upper* bound $t_{\rm f}^{\rm u}$ for the number of iterations. We find

$$t_{\rm f}^{\rm u}(\sigma) = 1 + \frac{2\arctan(1/2\sqrt{\varepsilon})}{\arctan(2\sqrt{\varepsilon})}. \tag{5.74}$$

And n_t reaching the value $\sigma \equiv \frac{1}{2} + \varepsilon$ gives a *lower* bound:

$$t_{\rm f}^{\rm l}(\sigma) = 1 + \frac{\arctan\left[\left(\frac{1}{4} - \varepsilon\right)/\sqrt{\varepsilon}\right] + \arctan(1/2\sqrt{\varepsilon})}{\arctan(2\sqrt{\varepsilon})}. \tag{5.75}$$

It is an easy exercise to show that

$$\arctan\left[(\frac{1}{4} - \varepsilon)/\sqrt{\varepsilon}\right] = \arctan\left(1/2\sqrt{\varepsilon}\right) - \arctan\left(2\sqrt{\varepsilon}\right), \tag{5.76}$$

which gives simply

$$t_{\rm f}^{\rm l} = t_{\rm f}^{\rm u} - 1. \tag{5.77}$$

The only feature that does not completely correspond to the breakdown of a physical bundle is that we have not ensured that the numbers N_t of unbroken fibers are *integers*. In the last stage with merely a handful of intact fibers, this could matter. We, therefore, check the analysis by simulation, as shown in Figure 5.5. The result is very satisfactory: the upper and lower bounds (5.74) and (5.75) nicely embrace the simulation results.

When the external load is large, just a few iterations suffice to induce complete bundle breakdown. On the other hand, when the external load approaches the critical load $\sigma_{\rm c} = 1/4$, the number of necessary iterations becomes very large. Near criticality ($\varepsilon \to 0$) both the upper and lower bounds, (5.74) and (5.75), have a square-root divergence:

$$t_{\rm f} \simeq \frac{1}{2}\pi(\sigma - \sigma_{\rm c})^{-1/2}, \tag{5.78}$$

to dominating order for small ε.

Before we leave the uniform threshold distribution, it is instructive to give the nonlinear dynamics explanation behind the critical divergence (5.78). Figure 5.6 shows n_{t+1} versus n_t, as given by the iteration (5.65) for $\sigma = 0.26$, slightly exceeding the critical stress value $\sigma_{\rm c} = 0.25$. The iteration path oscillates between the function $n_{t+1}(n_t)$ and the diagonal line $n_{t+1} = n_t$. Between the function and the diagonal line, there is a narrow tunnel in which the iterated values change slowly.

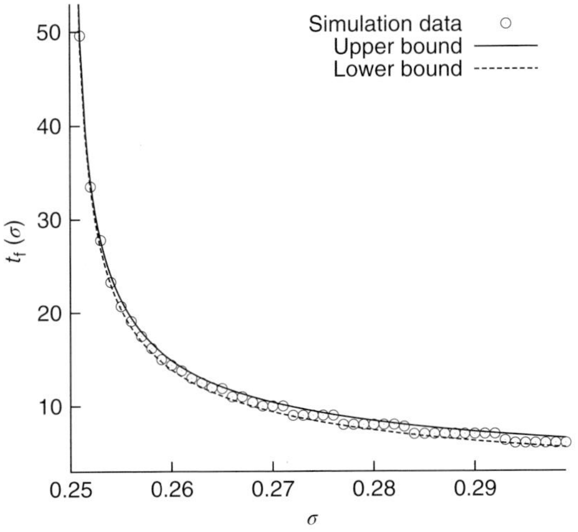

Figure 5.5 Simulation for the number of iterations t_f until every fiber is broken, for a bundle with the uniform threshold distribution on the unit interval. The simulation results, marked with asterisks, are averaged over 100 000 samples with $N = 10^6$ fibers in each bundle. The solid curve is the upper bound, Eq. (5.74), and the dashed curve is the lower bound, Eq. (5.75).

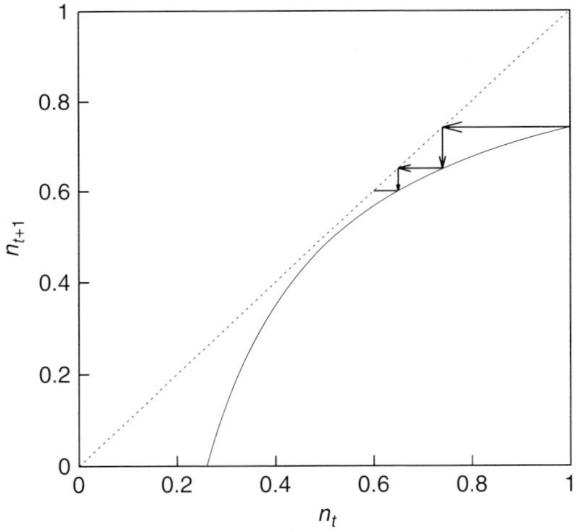

Figure 5.6 Iterations for the slightly supercritical uniform fiber strength distribution model ($\sigma = 0.26 = \sigma_c + 0.01$). The path of the iteration moves to and fro between the diagonal and the iteration function $n_{t+1} = 1 - \sigma/n_t$.

The tunnel contains many iterations, and when the external stress approaches the critical value, the tunnel width shrinks toward zero, and consequently, the number of iterations increases sharply. We argue that this will be the generic situation, the same for any threshold distribution.

5.2.2.2 General Threshold Distribution

We take the Weibull distribution (2.17) with index 5 (and $x_r = 1$), i.e.

$$P(x) = 1 - e^{-x^5} \tag{5.79}$$

as a typical example. The load curve $f(x) = x \exp(-x^5)$ gives a critical point x_c where $f'(x) = 0$, a maximal force equal to $\sigma_c = f(x_c)$, and by (5.6), the number of unbroken fibers at the critical point is $n_c = \sigma_c/x_c$. The actual values are

$$x_c = 5^{-1/5}, \quad \sigma_c = (5e)^{-1/5}, \quad \text{and} \quad n_c = e^{-1/5}. \tag{5.80}$$

Figure 5.7 shows, in analogy with Figure 5.6, the iterative function

$$n_{t+1} = 1 - P(\sigma/n_t) = e^{-(\sigma/n_t)^5}, \tag{5.81}$$

for the Weibull distribution with a slightly supercritical load and also the start of the iteration path. We concentrate our attention on the narrow channel where most of the iterations occur. The channel wall is almost parabolic and is well approximated by the second-order expression

$$n_{t+1} = n_t - a(n_t - n_c)^2 - b(\sigma - \sigma_c), \tag{5.82}$$

with constants a and b. Introducing $u_t = (n_t - n_c)/b$ and $\varepsilon = \sigma - \sigma_c$ we obtain

$$u_{t+1} - u_t = -Yu_t^2 - \varepsilon \quad \text{with} \quad Y = ab. \tag{5.83}$$

Since the iteration values u_t in the channel change very little with t, (5.83) is well approximated by the first-order ordinary differential equation

$$\frac{du}{dt} = -Yu^2 - \varepsilon. \tag{5.84}$$

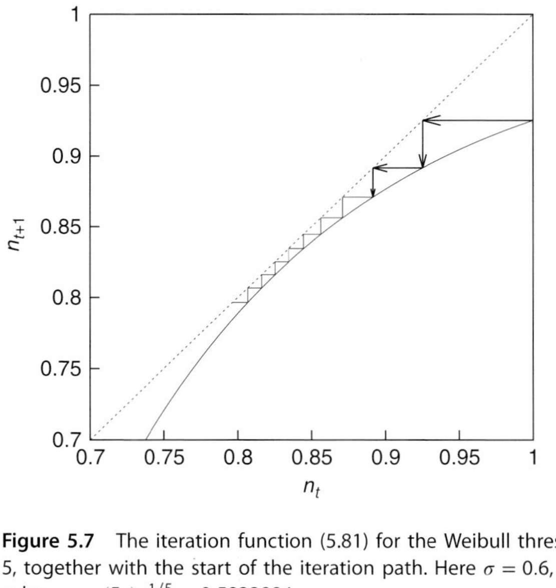

Figure 5.7 The iteration function (5.81) for the Weibull threshold distribution with index 5, together with the start of the iteration path. Here $\sigma = 0.6$, slightly larger than the critical value $\sigma_c = (5e)^{-1/5} = 0.5933994$.

Its solution,

$$\arctan(u\sqrt{Y/\varepsilon}) = -t\sqrt{Y\varepsilon} + \text{ constant}, \tag{5.85}$$

gives the number of iterations *in the channel*, starting with u_s and ending with u_e:

$$t_e - t_s = (Y\varepsilon)^{-1/2}[\arctan(u_s\sqrt{Y/\varepsilon}) - \arctan(u_e\sqrt{Y/\varepsilon})]. \tag{5.86}$$

Here, u_s is positive and u_e negative since $n_t - n_c$ is positive when the iteration path enters the channel and negative when it leaves. Therefore, the sharp parenthesis in (5.86) approaches the value π when $\varepsilon \to 0$. Consequently, the number of iteration steps in the channel has a square-root divergence near criticality,

$$t_e - t_s \simeq t_f \simeq \pi Y^{-1/2}\,(\sigma - \sigma_c)^{-1/2}. \tag{5.87}$$

This procedure for the near-critical behavior is valid for a general distribution of thresholds, while the value of the constant Y is case dependent. Therefore, the critical divergence

$$t_f \propto (\sigma - \sigma_c)^{-1/2} \tag{5.88}$$

is *universal*.

In order to determine the amplitude factor Y in (5.86) for a general threshold distribution $P(x)$, we must determine the constants a and b in the near-criticality expansion (5.82) of the iteration $n_{t+1} = 1 - P(\sigma/n_t)$. To first order in $\sigma - \sigma_c$ we have $P(\sigma/n_t) = P(\sigma_c/n_t) + (\sigma - \sigma_c)n_t^{-1}p(\sigma_c/n_t)$, so the iteration takes the form

$$n_{t+1} = 1 - P(\sigma_c/n_t) + (\sigma - \sigma_c)\,n_c^{-1}p(\sigma_c/n_c), \tag{5.89}$$

where n_t has been replaced by n_c in the small last term. It remains to expand $1 - P(\sigma_c/n_t)$ to second order in the small deviation $\Delta = n_t - n_c$:

$$P(\sigma_c/(n_c + \Delta)) = P(\sigma_c/n_c) - \Delta\,\sigma_c\,n_c^2 p(\sigma_c/n_c) \tag{5.90}$$

$$-\frac{1}{2}\Delta^2\left[p(\sigma_c/n_c)\sigma_c n_c^{-3} + \frac{1}{2}p'(\sigma_c/n_c)\,\sigma_c^2/n_c^4\right]. \tag{5.91}$$

A considerable simplification can be obtained by using $n_c = \sigma_c/x_c$, Eq. (5.6), and the criticality relation (2.15), $\sigma = x_c^2 p(x_c)$. The first term on the right-hand side equals n_c, and the second term equals $\Delta = n_t - n_c$. Altogether, one obtains

$$n_{t+1} = n_t - a(n_t - n_c)^2 - b(\sigma - \sigma_c), \tag{5.92}$$

with

$$a = [p(x_c) + x_c p'(x_c)/2]x_c^3/\sigma_c^2 \quad\text{and}\quad b = 1/x_c. \tag{5.93}$$

By (5.83) $Y = ab$, and the critical divergence (5.87) takes the form

$$t_f = K_+\,(\sigma - \sigma_c)^{-1/2}, \tag{5.94}$$

with the amplitude

$$K_+ = \pi\,\frac{2\sigma_c/x_c}{\sqrt{4p(x_c) + 2x_c p'(x_c)}}. \tag{5.95}$$

For the uniform distribution of thresholds, $p(x) = 1$, we recover the previous result (5.78).

5.2.3
Precritical Relaxation

We now consider external stresses less than the critical one, $\sigma < \sigma_c$, and use the positive parameter

$$\varepsilon = \sigma_c - \sigma \tag{5.96}$$

as a measure of the deviation from criticality. In this case, the bundle is expected to relax to an equilibrium situation with a finite number of intact fibers. The question is how many iteration t_f are needed to reach equilibrium.

5.2.3.1 Uniform Threshold Distribution

Again we start with the uniform threshold distribution (5.16), and again we transform the nonlinear iteration (5.20) to a linear one by means of two transformations. Introducing $\sigma = \frac{1}{4} - \varepsilon$ and

$$n_t = \frac{1}{2} + \sqrt{\varepsilon}\,/z_t \tag{5.97}$$

into (5.20), we have

$$2\sqrt{\varepsilon} = \frac{z_{t+1} - z_t}{1 - z_{t+1}z_t}. \tag{5.98}$$

A second transformation,

$$z_t = \tanh w_t, \tag{5.99}$$

gives

$$2\sqrt{\varepsilon} = \frac{\tanh w_{t+1} - \tanh w_t}{1 - \tanh w_{t+1} \tanh w_t} \equiv \tanh(w_{t+1} - w_t). \tag{5.100}$$

Hence we have the linear iteration $w_{t+1} - w_t = \operatorname{artanh}(2\sqrt{\varepsilon})$, which gives

$$w_t = w_0 + t \ \operatorname{artanh}(2\sqrt{\varepsilon}). \tag{5.101}$$

Here artanh denotes the inverse *function* of tanh. Written out, $\operatorname{artanh}(x) = \frac{1}{2} \ln\,[(1+x)/(1-x)]$. Via (5.97) and (5.99) the initial situation with no broken fibers, $n_0 = 1$, corresponds to $w_0 = \operatorname{artanh}(2\sqrt{\varepsilon})$, so that (5.101) becomes

$$w_t = (1+t) \ \operatorname{artanh}(2\sqrt{\varepsilon}). \tag{5.102}$$

For the original variable this corresponds to

$$n_t = \frac{1}{2} + \frac{\sqrt{\varepsilon}}{\tanh[(1+t)\operatorname{artanh}(2\sqrt{\varepsilon})]}. \tag{5.103}$$

After an infinite number of iterations ($t \to \infty$ in (5.103)) n_t apparently approaches the fixed point,

$$n^* = \frac{1}{2} + \sqrt{\varepsilon}, \tag{5.104}$$

which is the fixed point (5.18) for the uniform distribution. However, the bundle contains merely a finite number of fibers, so equilibrium should be reached after a *finite* number of steps. Since an equilibrium value corresponds to a fixed point, we seek fixed points N^* for finite N.

Taking into account that the variables N_t are integers, the iteration (5.4) for the uniform distribution may be written

$$N_{t+1} = N - [N^2\sigma/N_t], \tag{5.105}$$

where $[x]$ denotes the integer part of x. The corresponding fixed-point equation takes the form $N^* = N - [N^2\sigma/N^*]$. Since $x - 1 < [x] \leq x$, a fixed point N^* satisfies

$$N^2\sigma/N^* - 1 < N - N^* \leq N^2\sigma/N^*. \tag{5.106}$$

Replacing each inequality by an equality gives two quadratic equations for N^*. The first one, $N^2\sigma/N^* - 1 = N - N^*$, with $\sigma = \frac{1}{4} - \epsilon$, yields an *upper* critical value

$$N_{\rm u}^* = \frac{1}{2}(N+1) + \sqrt{N^2\epsilon + N + \frac{1}{4}} = N(\frac{1}{2} + \sqrt{\epsilon}) + \frac{1}{2} + \frac{1}{4}\epsilon^{-1/2} + \mathcal{O}(N^{-1}), \tag{5.107}$$

where a large-N expansion of the square root is used. The other equality, which corresponds to ignoring the integer restriction, clearly gives the critical value $N_{\rm l}^* = N(\frac{1}{2} + \sqrt{\epsilon})$, in accordance with (5.104).

Any integer in the interval $(N_{\rm l}^*, N_{\rm u}^*)$ will be a fixed point. Since the iterated values N_t are steadily decreasing, the first fixed point that is encountered will be the largest integer in the interval, close to $N_{\rm u}^*$ (see Ref. [67] for details). Using $n_t = N_{\rm u}^*/N$ as the final value in (5.103), we find the following estimate for the number of iterations necessary to reach this value,

$$t_{\rm f}(\sigma) = -1 + \frac{\operatorname{arcoth}[1 + (1 + 2\sqrt{\epsilon})/4N\epsilon]}{\operatorname{artanh}(2\sqrt{\epsilon})}. \tag{5.108}$$

It is not difficult to show that for small s we have $\operatorname{arcoth}(1+s) = \frac{1}{2}\ln\,(1 + 2/s) \simeq \frac{1}{2}\ln\,(2/s)$, so that for large N (5.108) can be simplified to

$$t_{\rm f}(\sigma) = -1 + \frac{\ln(N)}{2\,\operatorname{artanh}(2\sqrt{\epsilon})}. \tag{5.109}$$

The simulation data in Figure 5.8 for the uniform threshold distribution are well approximated by this analytic formula.

Equation (5.109) shows that near the critical point, the number of iterations has again a square-root divergence,

$$t_{\rm f} = K_-\,(\sigma_{\rm c} - \sigma)^{-1/2}, \tag{5.110}$$

now with a size-dependent amplitude,

$$K_- = \frac{1}{4}\,\ln(N). \tag{5.111}$$

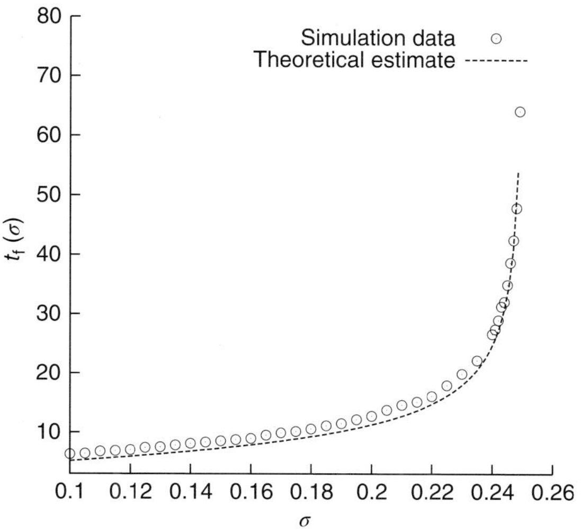

Figure 5.8 Simulation results for the number of iterations t_f necessary to reach equilibrium for the uniform threshold distribution (5.16). The graph is based on 10 000 samples with $N = 10^6$ fibers in each bundle. The dotted line is the theoretical estimate (5.109).

5.2.3.2 General Threshold Distribution

For a general fiber strength distribution, the iterated values n_t will, as we saw in Section 5.2.1.1, approach the fixed point with exponentially decreasing steps,

$$n_t - n^* \propto e^{-t/\tau}, \tag{5.112}$$

with a relaxation parameter

$$\tau = 1/\ln[n^{*2}\sigma/p(\sigma/n^*)]. \tag{5.113}$$

If we allow ourselves to use the exponential formula (5.112) all the way from the start at $n_0 = 1$, we obtain

$$n_t - n^* = (1 - n^*)\, e^{-t/\tau}. \tag{5.114}$$

The iteration stops after a finite number of steps, and it is a reasonable supposition to assume that it terminates when $N_t - N^*$ is of the order 1, that is, $n_t - n^* \simeq 1/N$. By (5.114) the number of iterations to reach this value is then

$$t_f = \tau \ \ln[N(1 - n^*)] \simeq \tau \ \ln(N) = \frac{\ln(N)}{\ln[n^{*2}\sigma^{-1}/p(\sigma/n^*)]}. \tag{5.115}$$

We have assumed N large and have kept merely the leading term.

The assumption that the exponential expression may be used from the outset of the iteration cannot always be defended. It works well in a *near-critical* situation, however, since almost all iterations occur near the fixed point, and the first few steps are of no importance.

For a Weibull fiber strength distribution with index 5, Eq. (5.79), the fixed-point Eq. (5.9) takes the form $n^* = \exp\,[-(\sigma/n^*)^5]$, i.e.

$$\sigma = n^*[-\ln(n^*)]^{1/5}. \tag{5.116}$$

For this distribution, (5.115) gives the total number of iterations equal to

$$t_{\mathrm{f}} = \frac{\ln(N)}{\ln[5(-\ln n^*)]}. \tag{5.117}$$

We have used that for this distribution $p(\sigma/n^*) = 5(\sigma/n^*)^4 \exp[-(\sigma/n^*)^5] = 5\sigma^4/n^{*3}$ and (5.116).

Equations (5.116) and (5.117) determine the number of iterations as a function of the stress, $t_{\mathrm{f}}(\sigma)$, and is compared with simulations in Figure 5.9. The agreement is satisfactory.

By using the result (5.61) for the relaxation parameter τ in (5.115) we have to dominating order

$$t_{\mathrm{f}} = K_{-}(\sigma_{\mathrm{c}} - \sigma)^{-1/2}, \tag{5.118}$$

now with a size-dependent amplitude K_{-},

$$K_{-} = \sigma_{\mathrm{c}} x_{\mathrm{c}}^{-1}[4p(x_{\mathrm{c}}) + 2x_{\mathrm{c}} p'(x_{\mathrm{c}})]^{-1/2} \ln(N). \tag{5.119}$$

For the uniform distribution of fiber strengths, $p(x) = 1$, we recover the previous result (5.111).

5.2.4
Critical Amplitudes

For large N, the derivations in Sections 5.2.2 and 5.2.3 have provided very good approximations to the total number t_{f} of relaxation steps, both in the precritical and the postcritical situations. Moreover, when the external stress σ approaches

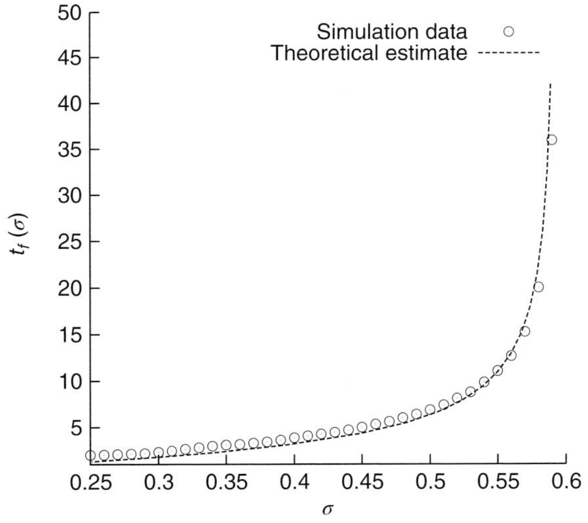

Figure 5.9 Simulation results for the number of iterations t_{f} necessary to reach equilibrium for the Weibull distribution with index 5. The graph is based on 1000 samples with $N = 10^6$ fibers in each bundle. The dashed line is the theoretical estimate (5.117).

the critical value σ_c from below or from above, it appears, from the nature of our procedures, that the results are asymptotically exact.

The asymptotic results for t_f show a square-root divergence at the critical point,

$$t_f(\sigma) = K_{\pm}\ |\sigma - \sigma_c|^{-1/2} \tag{5.120}$$

on both the precritical and the postcritical sides; see (5.94) and (5.118). This critical behavior is generally valid for all threshold distributions for fibers.

On the postcritical side, t_f is independent of the number N of fibers, as long as N is large. On the precritical side, however, the amplitude K_- is size dependent, proportional to the logarithm of N.

The ratio between the pre- and postcritical amplitudes (5.95) and (5.119),

$$\frac{K_-}{K_+} = \frac{\ln(N)}{2\pi}, \tag{5.121}$$

is universal [67].

In conclusion, in this chapter we have modeled the breakdown events in a fiber bundle as a stepwise process. The process is initiated by applying a finite external load to the bundle. In the first step, all fibers that cannot withstand the load fail, which gives an increased load on the remaining fibers and so on. The resulting iterative processes for effective stress and size of the intact bundle are interesting one-dimensional dynamical systems.

6
Predicting Failure

Predicting when failure occurs is a major challenge in various scenarios of fracture and breakdown – ranging from fracturing in nanomaterials to onset of earthquakes. A material body can tolerate a certain level of load or force, and beyond that level, it collapses. If the load is increased continuously, when does the collapse occur? Is there any precursor that signals that the complete failure is imminent?

In this chapter, we discuss how one can predict the failure point of a loaded fiber bundle from the available precursors. The term *precursor* usually means some prior indications of an upcoming incident, and in the current context, such incident is the complete failure (collapse) of a fiber bundle under external load. We start by considering equal-load-sharing models, in which the load previously carried by a failed fiber is shared equally by all the remaining intact fibers. The bundle consists of N elastic fibers, clamped at both ends (Figure 2.1). All the fibers obey Hooke's law with force constant κ set to unity for simplicity. Each fiber i is associated with a breakdown threshold x_i for its elongation. When the length exceeds x_i the fiber breaks immediately and does not contribute to the strength of the bundle thereafter. The individual thresholds x_i are assumed to be independent random variables with the same cumulative distribution function $P(x)$ and a corresponding density function $p(x)$.

In the first section, we consider situations in which the distribution of some quantity changes character when failure is imminent [69]. The change of character is seen as a crossover from one type of asymptotic behavior to another. In the second section, we point out that in some cases the failure point can be estimated by extrapolation. In the third section, we consider situations in which the bundle is overloaded, with the aim both to have a signal that shows that it is overloaded and to have some indication of when the complete breakdown will occur.

For prediction purpose, it is important that precursors can be seen in a single system. In addition to the average behavior, we, therefore, present and discuss results that can be seen in a single bundle containing a large number of fibers.

Similar crossover phenomena can be seen in other systems. In Section 6.1.3.2, we give a short account of one such system, the *fuse model*, a two-dimensional network of electrical fuses.

The Fiber Bundle Model: Modeling Failure in Materials, First Edition.
Alex Hansen, Per C. Hemmer and Srutarshi Pradhan.

6.1 Crossover Phenomena

In strain-controlled loading, at each step the whole bundle is stretched until the weakest fiber among the intact ones fails. At an elongation x, the total force on the bundle is x times the number of *intact* fibers. We showed in Section 2.3 that the expected or average force at this stage is

$$F(x) = N\,x\,(1 - P(x)), \tag{6.1}$$

and that the maximum of $F(x)$ corresponds to the critical value x_c, satisfying

$$1 - P(x_c) - x_c p(x_c) = 0. \tag{6.2}$$

Elongations beyond the critical one cause bundle collapse.

6.1.1 The Avalanche Size Distribution

For a bundle controlled by the external load, a fiber failure results in an increased stress on the remaining intact fibers. This may in turn trigger further fiber failures, which produce bursts (avalanches) that lead either to a stable situation or to breakdown of the whole bundle. A burst can be characterized by the number Δ of fibers that fail simultaneously. It was shown in Chapter 3 that when all bursts are recorded for the entire failure process, the average number of burst events of size Δ follows an asymptotic power law of the form

$$\frac{D(\Delta)}{N} = C\Delta^{-\xi} \tag{6.3}$$

in the limit $N \to \infty$. Here, $\xi = 5/2$ is the *universal* burst exponent and C is a constant. Universality means that the value of ξ is, under a mild assumption, independent of the threshold distribution $P(x)$: We assume that the load curve, the average force $F(x)$, has a quadratic maximum.

However, if we just sample bursts that occur near the breakdown point x_c, a different power law is seen. In Figure 6.1 we compare the complete burst distribution (6.3) with what we get when we sample merely bursts in the near-critical strain interval $(0.9x_c, x_c)$. Two very different burst distributions are seen. This result will now be explained analytically.

In Chapter 3 we provided a complete analytic formulation (3.52) for the distribution of bursts starting somewhere in the interval (x_0, x_c):

$$\frac{D(\Delta)}{N} = \int_{x_0}^{x_c} \frac{\Delta^{\Delta-1}}{\Delta!}[a(x)e^{-a(x)}]^{\Delta} a(x)^{-1}[1 - a(x)]p(x)\,\mathrm{d}x, \tag{6.4}$$

where $a(x)$ is given by (3.53), $a(x) = xp(x)/[1 - P(x)]$. From Eq. (6.2), we see that $a(x) = 1$ at the critical point x_c. Since ae^{-a} is maximal for $a = 1$, which corresponds to the critical point, we expand around $x = x_c$. This has already been done in

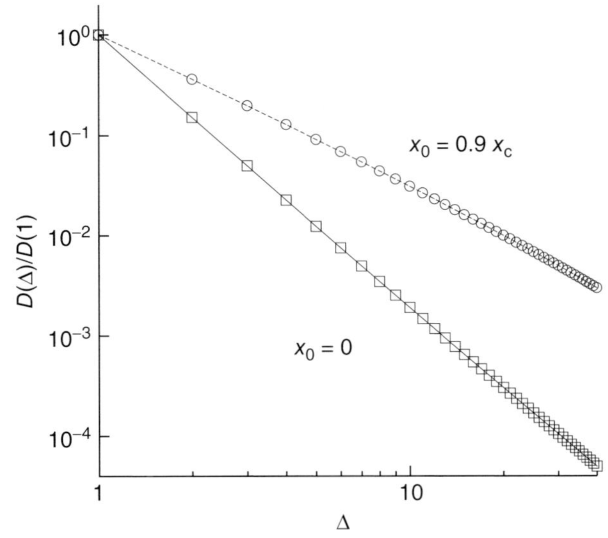

Figure 6.1 The distributions of bursts in the fiber bundle model with $x_0 = 0$ and $x_0 = 0.9x_c$. The figure is based on 10 000 samples with $N = 10^6$ fibers with uniformly distributed fiber thresholds between 0 and 1.

Section 3.3.2, Eq. (3.68), with the result

$$\frac{D(\Delta)}{N} = \frac{\Delta^{\Delta-2}\, e^{-\Delta} p(x_c)}{|a'(x_c)|\Delta!} [e^{-\Delta\, a'(x_c)^2(x-x_c)^2/2}]_{x_0}^{x_c} \tag{6.5}$$

By use of the Stirling approximation $\Delta! \simeq \Delta^\Delta e^{-\Delta}\sqrt{2\pi\Delta}$, the burst distribution (6.5) may be written as

$$\frac{D(\Delta)}{N} = C\Delta^{-5/2}(1 - e^{-\Delta/\Delta_c}) = \begin{cases} C\Delta_c^{-1}\,\Delta^{-3/2} & \text{for } \Delta \ll \Delta_c \\ C\Delta^{-5/2} & \text{for } \Delta \gg \Delta_c \end{cases} \tag{6.6}$$

with a constant

$$C = (2\pi)^{-1/2} p(x_c)/|a'(x_c)| \tag{6.7}$$

and with

$$\Delta_c = \frac{2}{[a'(x_c)(x_c - x_0)]^2}. \tag{6.8}$$

We see from (6.6) that there is a crossover at a burst length of the order of Δ_c:

$$\frac{D(\Delta)}{N} \propto \begin{cases} \Delta^{-3/2} & \text{for } \Delta \ll \Delta_c \\ \Delta^{-5/2} & \text{for } \Delta \gg \Delta_c \end{cases} \tag{6.9}$$

When the sampling interval is very close to criticality, as in Figure 6.1, Δ_c is very large, and the $\Delta^{-5/2}$ behavior is hardly seen.

We have thus shown the existence of a crossover from the generic asymptotic behavior $D \propto \Delta^{-5/2}$ to the power law $D \propto \Delta^{-3/2}$ near criticality, that is, near global breakdown. The crossover is a universal phenomenon, independent of the threshold distribution $p(x)$.

For, the uniform distribution $\Delta_c = (1 - x_0/x_c)^{-2}/2$, so for $x_0 = 0.8\,x_c$, we have $\Delta_c = 12.5$. For the Weibull distribution $P(x) = 1 - \exp\,(-(x-1)^{10})$, where $1 \le x \le \infty$, we get $x_c = 1.72858$, and for $x_0 = 1.7$, the crossover point will be at $\Delta_c \simeq 14.6$. Such crossover is clearly observed (Figure 6.2) near the expected values $\Delta = \Delta_c = 12.5$ and $\Delta = \Delta_c = 14.6$, respectively, for these threshold distributions. Note that the crossover is first seen in the *small* burst region.

The simulation results shown so far are based on *averaging* over a large number of samples. For applications, it is important that the crossover signal can also be seen in a single sample. We show in Figure 6.3 that equally clear crossover behavior is seen in a *single* fiber bundle when N is large enough. Also, as a practical tool, one must sample finite intervals (x_i, x_f) during the fracture process. The crossover will be observed when the interval is close to the failure point. Figure 6.3 compares the complete burst distribution with what we get when we sample merely bursts in the threshold interval $(0.9x_c, x_c)$.

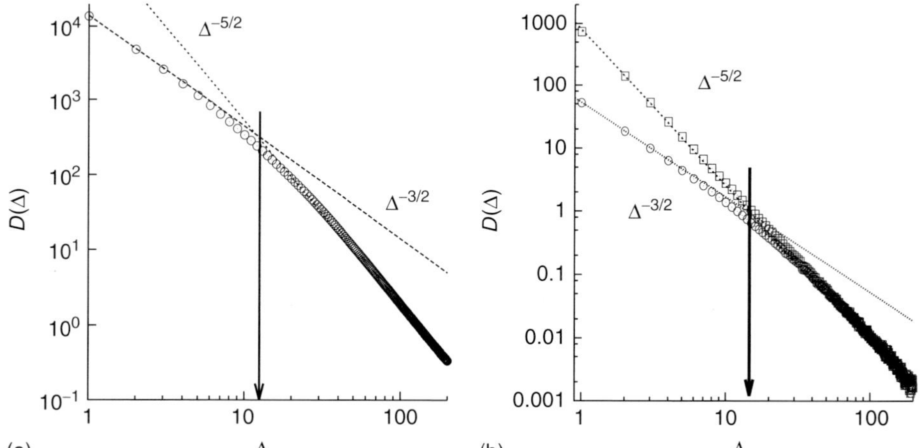

Figure 6.2 The distribution of bursts for the uniform threshold distribution (a) with $x_0 = 0.80x_c$ and for a Weibull distribution of index 10 (b) with $x_0 = 1$ (square) and $x_0 = 1.7$ (circle). Both the figures are based on 50 000 samples with $N = 10^6$ fibers each. The straight lines represent two different power laws, and the arrows indicate the crossover points $\Delta_c \simeq 12.5$ and $\Delta_c \simeq 14.6$, respectively.

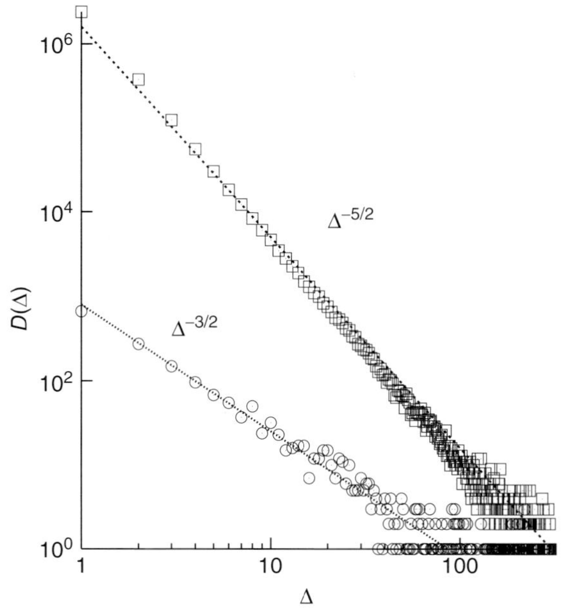

Figure 6.3 Burst distributions: all bursts (squares) and bursts within an interval $(0.9x_c, x_c)$ (circles). The figure is based on a single bundle containing $N = 10^7$ fibers with uniformly distributed fiber thresholds between 0 and 1.

6.1.1.1 Burst Avalanches at Criticality

At criticality, we have by Eq. (6.5) $\Delta_c = \infty$, and the 5/2 power law is no longer present. We can substantiate this by using a random walk representation to demonstrate that at criticality the asymptotic burst distribution follows a 3/2 power law.

In Section 3.3.5.1, we derived the exact random walk for the force steps $\Delta F = F_{k+1} - F_k$. At criticality, the probability $\rho(f)\,df$ that ΔF lies in the interval $(f, f + df)$ is given by (3.98),

$$\rho(f) = \begin{cases} x_c^{-1}\, e^{-1-f/x_c} & \text{for } f \geq -x_c \\ 0 & \text{for } f < -x_c \end{cases} \tag{6.10}$$

We have used that at criticality $1 - P(x_c) = x_c p(x_c)$, Eq. (6.2). This random walk is unsymmetrical, but unbiased, $\langle f \rangle = 0$, as it should be at criticality.

The first burst of size Δ corresponds to a random walk in which the position after each of the first $\Delta - 1$ steps is lower than the starting point, but after step number Δ, the position of the walker exceeds the starting point. The probability of this equals

$$\begin{aligned}\mathrm{Prob}(\Delta) &= \int_{-x_c}^{0} \rho(f_1)\mathrm{d}f_1 \int_{-x_c}^{-f_1} \rho(f_2)\mathrm{d}f_2 \int_{-x_c}^{-f_1-f_2} \rho(f_3)\mathrm{d}f_3 \ldots \\ &\ldots \int_{-x_c}^{-f_1-f_2\cdots-f_{\Delta-2}} \rho(f_{\Delta-1})\mathrm{d}f_{\Delta-1} \\ &\int_{-f_1-f_2\cdots-f_{\Delta-1}}^{\infty} \rho(f_\Delta)\,\mathrm{d}f_\Delta. \end{aligned} \tag{6.11}$$

The last integral is easy. By means of (6.10) we have

$$\int_{-f_1-f_2\cdots-f_{\Delta-1}}^{\infty} \rho(f_\Delta)\mathrm{d}f_\Delta = e^{-1}\, e^{(f_1+f_2+\,\ldots\,+f_{\Delta-1})/x_c}. \tag{6.12}$$

Since $\rho(f)e^{f/x_c} = e^{-1}/x_c$ we end up with

$$\begin{aligned}\mathrm{Prob}(\Delta) = e^{-\Delta} \int_{-1}^{0} \mathrm{d}f_1 \int_{-1}^{-f_1} \mathrm{d}f_2 \int_{-1}^{-f_1-f_2} \mathrm{d}f_3 \\ \ldots \int_{-1}^{-f_1-f_2\,\ldots\,-f_{\Delta-2}} \mathrm{d}f_{\Delta-1}. \end{aligned} \tag{6.13}$$

For simplicity, we have put $x_c = 1$, since the quantity x_c merely determines the scale of the steps, and here it is only the relative step lengths that matter.

In Appendix A.4, the expression (6.13) is evaluated, with the result

$$\mathrm{Prob}(\Delta) = \frac{e^{-\Delta}\,\Delta^{\Delta-1}}{\Delta!} \simeq \frac{1}{\sqrt{2\pi}}\,\Delta^{-3/2}. \tag{6.14}$$

The simulation results in Figure 6.4 are in excellent agreement with this distribution. At the completion of a burst, the force, that is, the excursion of the random walk, is larger than all previous values. Therefore, this point may be used as a new starting point to find, by the same calculation, the distribution of the next burst and so on. Consequently, the complete burst distribution at criticality is essentially proportional to $\Delta^{-3/2}$.

6.1.2
Energy Bursts

We saw in Section 3.3.6 that bursts with a mechanical energy release E had at high energies an asymptotic density

$$g(E) \propto E^{-5/2}. \tag{6.15}$$

Again, if one merely samples energy bursts that occur near bundle breakdown, a different asymptotics will be seen. To see that we go back to the high-energy expression (3.129) for $g(E)$,

$$\frac{g(E)}{N} = E^{-5/2}\frac{[2p(x_c)+x_c p'(x_c)]x_c^2}{4\sqrt{\pi}a'(x_c)^2}\left[\exp\left[-E(x_c-x)^2 a'(x_c)^2/x_c^2\right]\right]_{x_0}^{x_c}. \tag{6.16}$$

Here we have assumed that the sampling corresponds to strain values in the interval (x_0, x_c), instead of the complete interval $(0, x_c)$ that was used in Section 3.3.6.1.

Since

$$\left[\exp\left[-E(x_c - x)^2 a'(x_c)^2/x_c^2\right]\right]_{x_0}^{x_c} = 1 - e^{-E/E_c}, \tag{6.17}$$

with

$$E_c = \frac{x_c^2}{a'(x_c)^2(x_c - x_0)^2}, \tag{6.18}$$

there is in this case a crossover near the energy E_c:

$$\frac{g(E)}{N} \propto \begin{cases} E^{-3/2} & \text{for} \quad E \ll E_c \\ E^{-5/2} & \text{for} \quad E \gg E_c \end{cases} \tag{6.19}$$

Thus, the energy distribution in bursts (Figure 6.5) exhibits a similar crossover behavior as the avalanche size distribution.

In conclusion, we have for the fiber bundles, both for the burst size and for the energy burst magnitude, demonstrated a crossover behavior with two different power laws with exponents −3/2 and −5/2. This crossover signals that catastrophic failure is imminent.

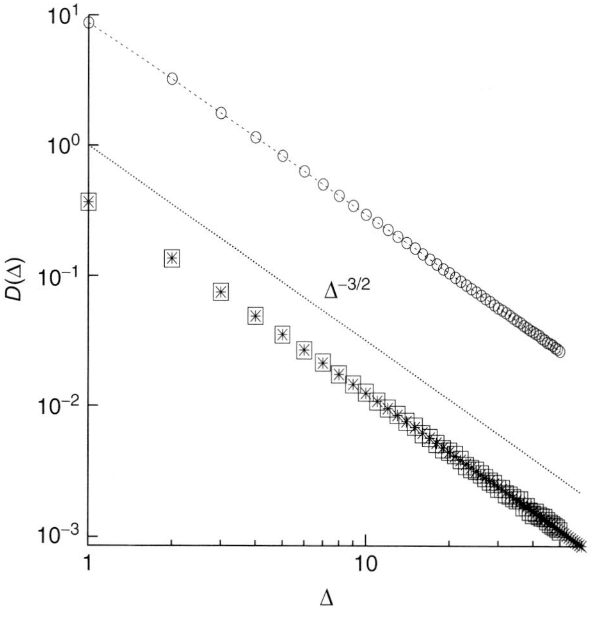

Figure 6.4 Distribution of first bursts (squares) and total bursts (circles) for the critical strength distribution with $x_0 = x_c$. The simulation results are based on 10^6 samples with $N = 80\,000$ fibers. The "star" symbol stands for the analytic result (6.14).

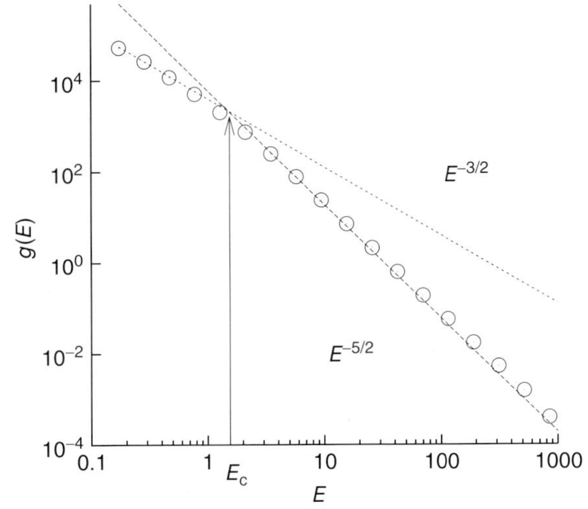

Figure 6.5 Distribution of energy bursts (circles) for the uniform strength distribution with $x_0 = 0.9x_c$. The simulation results are based on 10^5 samples with $N = 10^6$ fibers. The "arrow" indicates the crossover energy magnitude E_c, which follows the analytic result (6.18).

6.1.3 The Crossover Phenomenon in Other Systems

The observation of a crossover in avalanche distributions may be of practical importance, as it gives a criterion for the imminence of complete failure. The bursts or avalanches can be recorded from outside – without disturbing the ongoing failure process. Therefore, any signature in burst statistics that can warn of imminent system failure would be very useful and have a wide range of applicability. It is enticing to note the recent observation of a crossover behavior in the magnitude distribution of earthquakes before the largest earthquake appears. A similar crossover behavior is also seen in the burst distribution and energy distribution of the fuse model, which is a standard model for studying fracture and breakdown phenomena in disordered systems. Most important is that this crossover signal does not hinge on observing rare events and is seen in a single system (see Figure 6.3). Therefore, such a crossover signature has strong potential as a detection tool.

6.1.3.1 Earthquakes

The Gutenberg–Richter law correlates the magnitude and frequency of earthquakes in a linear way [70]. If Ω is the frequency (i.e., number of earthquakes per time unit) and M is the magnitude, the correlation reads

$$\log_{10}\Omega = a - bM, \tag{6.20}$$

where a and b are constants. The coefficient a varies from region to region and measures how prone that region is. The coefficient b measures the relative frequencies of earthquakes and is insensitive to where it is measured. It is typically in the range 0.8–1.2. Much effort has gone into understanding whether – and if so, why – b has a universal value [71]. In 1981, Smith suggested the use of time variations in the b value as an earthquake predictor [72]. Based on the data from New Zealand, he measured b by using a sliding 50-event window in the time interval 1955–1979 and $M \geq 4.0$. His analysis indicated that before large events, the b value would start to increase, indicating an increase in the frequency of small-scale events compared to large-scale events. The time at which the increase sets in varied substantially from 6 months to 7 years. Close to the large event, the b value returns to the normal level. Latchman *et al.* [73] report similar observations.

A number of recent studies [74-77] have focused on the seismic activity before large earthquakes and show a drop in the b value in the window right before the main shock, as reported by Kawamura [74]. See Figure 6.6. Here, the b value for Japanese earthquakes, 0.88, drops to $b = 0.60$ before large events.

What does this have to do with the topic of this chapter? The Richter magnitude scale M was originally defined with respect to a particular response of Dr. Richter's seismograph situated at Caltech. However, it turns out to be proportional to the logarithm of the elastic energy released during the earthquake, $M \propto \log_{10} E_e$. Hence, the Gutenberg–Richter law is in fact a power law relating

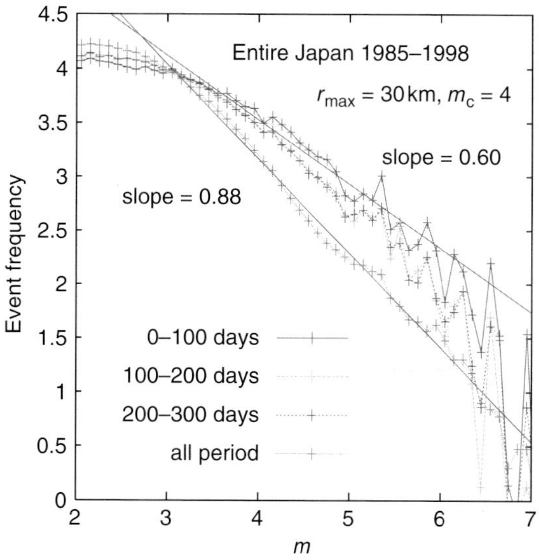

Figure 6.6 Crossover signature in the local magnitude (m) distributions of earthquakes in Japan. The exponent of the distribution during 100 days before a main shock is about 0.60, much smaller than the average value 0.88 [74].

frequency of events to their energy release,

$$\Omega = A\, E^{-b}. \tag{6.21}$$

Hence, cast in this form, we see the relation between the Gutenberg–Richter law and the energy burst distribution discussed in the previous section. It is tempting to see the change of b before large events, for example, the change of $b = 0.88 \rightarrow 0.6$ seen by Kawamura, as related to the change of exponent seen in the energy burst statistics seen in the fiber bundle model near catastrophic failure.

Should one get excited, is this a door to earthquake *prediction*? That is too much to expect. There is a strong stochastic element here and one has to settle with this being a step in the direction of earthquake *forecasting*.

Charles Richter and Beno Gutenberg

Courtesy of Caltech Seismological Laboratory, USA

Charles Francis Richter was born in 1900 in Overpeck, Ohio, to a twice married and soon twice divorced couple. He saw his father just once and took the family name of his mother, Lillian Anna Richter. They moved to Los Angeles. Richter's early scientific interests were not directed downward to the earth but upward. He was a competent amateur astronomer and got his undergraduate degree in physics at Stanford. He was a very talented but also troubled student and had to spend most of the year after graduation in a sanatorium.

Recovered, he turned to the California Institute of Technology for graduate studies. At Caltech, the Nobel laureate Robert Millikan was the star. On Millikan's suggestion, Richter began work on his PhD in theoretical physics, on the puzzling behavior of electron states in the hydrogen atom. After his PhD in 1928, Richter was summoned by Millikan. He was told that the nascent Seismological Laboratory wanted to hire a part-time research assistant with a background in physics. Was he interested? Although he had no intention of leaving theoretical physics, Richter accepted the job and started working with seismometry. Millikan found it necessary to strengthen the scientific staff at the laboratory and hired Beno Gutenberg in 1930.

Courtesy of Caltech Seismological Laboratory, USA

Beno Gutenberg was born 1889 in Darmstadt, Germany, where his father owned a small soap factory. He studied physics and mathematics at the University of Göttingen with Born, Hilbert, Klein, Minkowski, Prandtl, K. Schwarzschild, Voight and Weyl. He obtained his doctorate on a thesis of microseisms with Emil Wiechert, a physicist who became the world's first professor of geophysics. Microseisms are small disturbances more or less continuously recorded by sensitive seismometers. Gutenberg visited in 1910 the coast of Norway and was able to correlate surf in Norway with microseisms in Göttingen!

Gutenberg had a position at the University of Strasbourg, which he lost when Strasbourg became French in 1918. His applications for positions at German universities were unsuccessful, so he had to depend on a job in his father's factory. He continued his research in his spare time, published extensively, and soon became one of the world's leading seismologists. Probably Gutenberg's difficulties with obtaining an academic position in Germany was not unrelated to his Jewish background, and in any case he would, as most Jewish scientists in Germany, have lost such a position in 1933. The offer from Millikan in 1930 was, therefore, very welcome. The offer included a position as Professor of Geophysics at Caltech and directorship of the Seismological Laboratory.

When Gutenberg arrived at Caltech, his drive and organizational skills infused the young Earth science department and the Seismological Laboratory with energy, and soon the laboratory became the world's leading seismological institution. Gutenberg teamed up with the local scientists, in particular with Richter, and a long and fruitful collaboration between the two men started. Their development of a rating scale to measure the relative sizes of earthquake sources, known as the Richter's scale, is very well known. It would seem natural, in the opinion of many of their colleagues, to have also Gutenberg's name attached to the scale, but the name *Richter's scale* stuck. One reason is that Richter popularized the concept in newspapers and radio, since Gutenberg was hesitant to be interviewed due to his hearing problem. Another well-known joint work was the *Gutenberg–Richter law*, which expresses the relationship between the magnitude and the total number of earthquakes of that magnitude in any given region.

The personalities of the two men were very different. Gutenberg was personably and lively, played the piano, and had chamber music evenings at home. Albert Einstein was his friend and played the violin in one such happening. Richter, however, was awkward, and social interactions were troublesome for him. (His biographer, S. E. Hough, suggests that he suffered from the Asperger syndrome.) Nevertheless, they cooperated very well until Gutenberg passed away in 1960, wrote several monumental papers and a book, *Seismicity of the Earth*, together. Richter died in 1985.

6.1.3.2 The Fuse Model

At a less speculative level than the possible tie to earthquake statistics, is the crossover phenomenon seen in the fuse model near breakdown. The fuse model was introduced in 1985 by de Arcangelis *et al.* [78] Since then, hundreds of papers have been written on this model and we point to Alava *et al.* [79] for a review. The model consists of a network of electrical fuses as seen in Figure 6.7. Each bond in the lattice behaves as an ohmic resistor up to a maximum current t_{ij}, where i and j are neighboring nodes. If this threshold t_{ij} is exceeded, the bond burns out and the conductance of the bond drops to zero irreversibly. The thresholds are drawn from a statistical distribution, which is normally taken to be spatially uncorrelated. An increasing current is then driven through the lattice and the successive breakdown of the fuses is recorded.

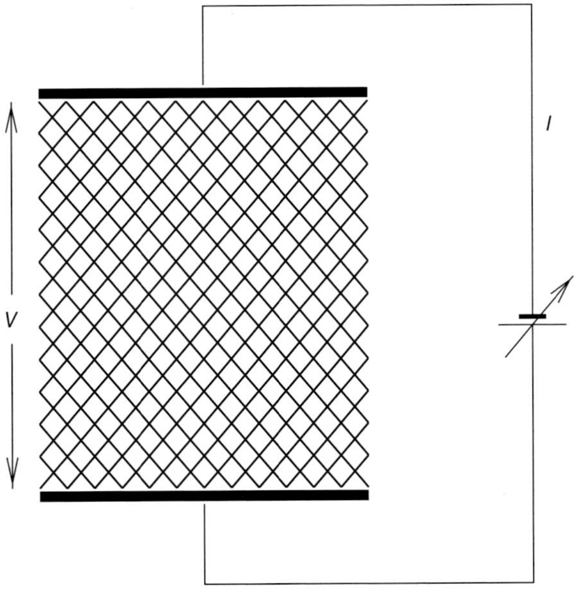

Figure 6.7 Each bond in this network is a fuse. A voltage V is applied across the network, resulting in a corresponding current I. As V is increased, the fuses will burn out one by one.

The fuse network is initiated with all intact fuses. The voltage V across the network is then increased gradually and the fuse with the smallest threshold t burns out. The current redistributes itself instantaneously in the network. The voltage is then further gradually increased until a second fuse burns out and so on.

The process of "gradually increasing the voltage across the network" is *not* implemented literally on the computer. Rather, the breakdown process is implemented as follows. First, note that for the range of voltage differences V that does not lead to breakdown of fuses – that is, the voltages that lie between two consecutive breakdowns – the voltage–current relation is *linear*. The network conductance is G, so that $I = GV$. The voltage difference between the neighboring nodes i and j is v_{ij}, and the corresponding current is $I_{ij} = g_{ij}v_{ij}$, where g_{ij} is the conductance of the bond. The total power dissipated in the network is

$$E_{\rm d} = VI = GV^2 = \frac{1}{2} \sum_{ij} g_{ij} v_{ij}^2 \,. \tag{6.22}$$

Thus, if we scale the voltage across the network $V \to \lambda V$, then the voltage across bond ij is scaled $v_{ij} \to \lambda v_{ij}$. Set the voltage across the network equal to 1. Identify the bond for which v_{ij}/t_{ij} is maximum and set

$$\frac{1}{\lambda} = \max_{ij} \frac{v_{ij}}{t_{ij}}. \tag{6.23}$$

Thus, if we scale $V = 1 \to V = \lambda$, we find

$$\max_{ij} \frac{\lambda v_{ij}}{t_{ij}} = \lambda \cdot \frac{1}{\lambda} = 1, \tag{6.24}$$

which is the maximum voltage difference the bond that is to break down next can sustain. Therefore, $V = \lambda$ is the voltage difference across the network at which the fuse blows. The breakdown algorithm is, therefore,

1) Calculate all v_{ij} given $V = 1$.
2) Identify $\max_{ij}(v_{ij}/t_{ij}) = 1/\lambda$.
3) Remove the bond with the maximum ratio.
4) Test whether the network still conducts. If yes, go to 1, otherwise stop.

This is essentially the same procedure as was discussed in Section 4.1.2.

Once λ is known, we have the current the network carries when that particular fuse blows from Eq. (6.22), $I = G\lambda$.

The breakdown process in the fuse model may proceed in a number of different ways, depending on the behavior of the threshold probability density, $p(t)$, for small and for large t. Early on in the breakdown process, starting from an intact lattice, the fuses that blow do so because their thresholds are small. Then, as holes appear in the lattice, fuses start blowing because they carry large currents. Hence, early on in the process, it is "disorder driven," and later on it is current driven.

Bursts may be recorded in the fuse model in a similar way as is done in the fiber bundle model [80]. The voltage drop across the fuse network, V, corresponds to the elongation of the fiber bundle, whereas the current that the fuse network carries, I, corresponds to the total force on the fiber bundle. At the point when fuse number k is about to fail, the minimum current flowing in the network for this to happen is I_k. The corresponding voltage across the network is V_k. We may now define bursts precisely in the same way as was done in the fiber bundle model: a burst of size Δ starting when fuse number k fails while controlling the current I would be defined as

$$I_k > I_{k-1} , \tag{6.25}$$

$$I_{k+j} \leq I_k \quad \text{for} \quad 1 \leq j < \Delta - 1 , \tag{6.26}$$

and

$$I_{k+\Delta} > I_k . \tag{6.27}$$

We show in Figure 6.8 the burst distribution in the fuse model based on 300 lattices of size 100 × 100 with a uniform threshold distribution on the unit interval. On average, catastrophic failure sets in when 2097 fuses have blown. The squares denote the burst distribution based on bursts appearing after the first 1000 fuses have blown. The triangles denote the burst distribution after 2090 fuses have blown. The two straight lines indicate power laws

$$D(\Delta) \sim \Delta^{-3} \tag{6.28}$$

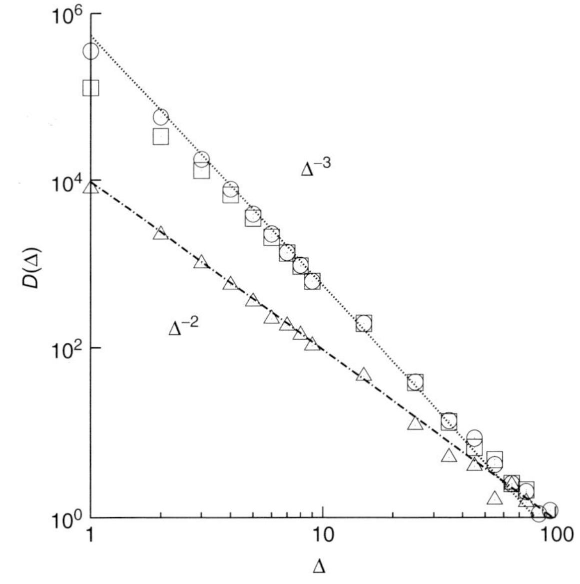

Figure 6.8 The burst distribution in the fuse model based on 300 fuse networks of size 100 × 100 using a uniform threshold distribution on the unit interval. The circles denote the burst distribution measured throughout the entire breakdown process. The squares denote the burst distribution based on bursts appearing after the first 1000 fuses have blown. The triangles denote the burst distribution after 2090 fuses have blown.

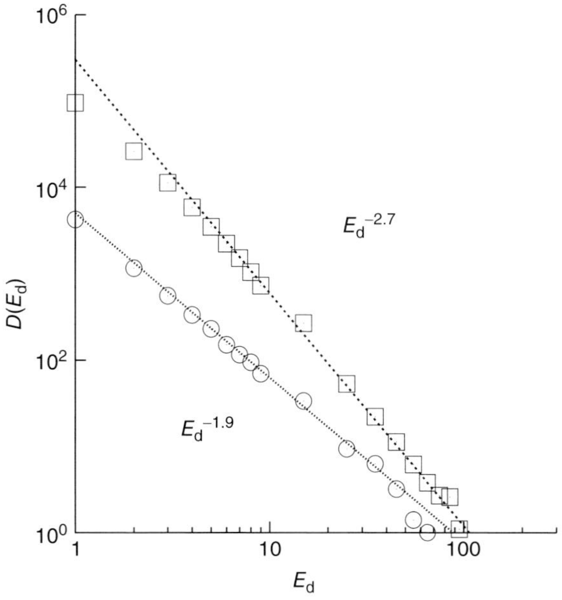

Figure 6.9 The power dissipation avalanche histogram $D(E_d)$ in the fuse model. The slopes of the two straight lines are −2.7 and −1.9, respectively. The squares show the histogram of avalanches recorded through the entire process, whereas the circles show the histogram recorded only after 2090 fuses have blown. The system size and the number of samples are the same as in Figure 6.8.

and

$$D(\Delta) \sim \Delta^{-2} , \tag{6.29}$$

respectively. Our conclusion is that we see the same effect as in the fiber bundle model: when the bursts are recorded away from (and all the way up to) the point at which the system fails, the bursts follow a power-law distribution with a large exponent, the value being 3.0. When bursts are recorded only near the failure point, the exponent is appreciably smaller; it is 2.0.

The power dissipation in the network also shows a crossover phenomenon. In Figure 6.9, we record the avalanche distribution for power dissipation, $D(E_d)$. Recording, as before, the avalanche distribution throughout the entire process and recording only close to the point at which the system catastrophically fails result in two power laws, with exponents $\xi = 2.7$ and $\xi = 1.9$, respectively. It is interesting to note that in this case there is not a difference of unity between the two exponents.

The power dissipation in the fuse model corresponds to the stored elastic energy in a network of elastic elements. Hence, the power dissipation avalanche histogram would in the mechanical system correspond to the released elastic energy. Such a mechanical system would serve as a simple model for earthquakes. The Gutenberg–Richter law relating the frequency of earthquakes with their magnitude is essentially a measure of the elastic energy released in the earth's crust, as the magnitude of an earthquake is the logarithm of the elastic energy released. Hence, the power dissipation avalanche histogram $D(E_d)$ in the fuse model is the quantity corresponding to the quantity that the Gutenberg–Richter law addresses in seismology. Furthermore, the power-law character of $D(E_d)$ is consistent with the form of the Gutenberg–Richter law. It is then intriguing that there is a change in exponent ξ also for this quantity when failure is imminent.

Even though nobody so far has been able to derive these results for the fuse model, they point to a similar underlying mechanism as that seen in the equal-load-sharing fiber bundle model. This hints at the results derived in this chapter for the equal-load-sharing model are more general than they seem.

6.2 Variation of Average Burst Size

We have seen that if the external load is increased quasistatically on a bundle with a large number of fibers, bursts of different sizes occur during the whole breaking process till complete failure. One can ask – what is the average burst size at a particular elongation value x? The average burst size is indeed a very relevant quantity that can be measured easily during the failure process. In Chapter 3, we have calculated in Eq. (3.64) the average burst size at an elongation x as

$$\langle \Delta \rangle = \frac{1}{1 - a(x)} = \frac{1 - P(x)}{1 - P(x) - xp(x)}, \tag{6.30}$$

which shows that the average burst size diverges at the critical point. For a uniform distribution of fiber thresholds ($P(x) = x$ for $0 \leq x \leq 1$), we get

$$\langle \Delta \rangle = \frac{1 - x}{1 - 2x}. \tag{6.31}$$

The theoretical estimate matches well with numerical observations (Figure 6.10). Now if we plot the inverse of the average burst size versus x, we should get a curve that meets the x axis at $x = x_c$, thus predicting the breakdown point. Even in a single system, we can see this signature (Figure 6.11).

6.3 Failure Dynamics Under Force-Controlled Loading

In the force-controlled case, the load is increased on a bundle at each loading step. Here, the failure dynamics of the bundle can be represented by a recursion relation. The entire recursive dynamics has been discussed in Chapter 5. We observed that both susceptibility and relaxation time diverge at the critical point following well-defined power laws. We recall near-critical variation of susceptibility and relaxation time with applied stress value (Eqs. (5.24) and (5.27))

$$\chi = \frac{1}{2}\,(\sigma_c - \sigma)^{-\beta} \text{ with } \beta = \frac{1}{2}. \tag{6.32}$$

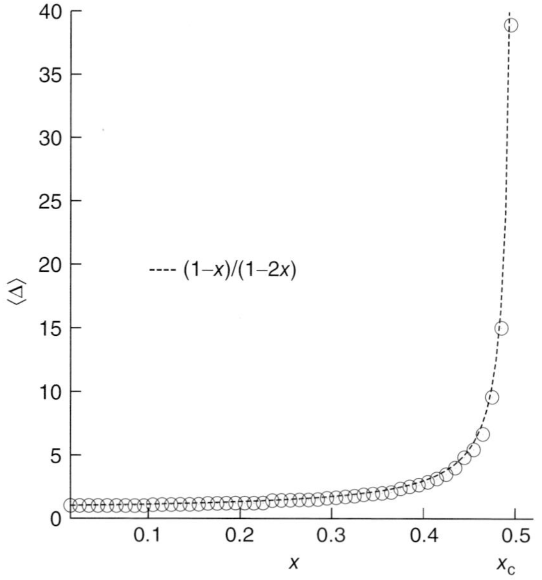

Figure 6.10 Simulation results for the average burst size Δ versus elongation x for a single bundle of 10^7 fibers with thresholds uniformly distributed in the unit interval. The solid line is the theoretical result (6.31).

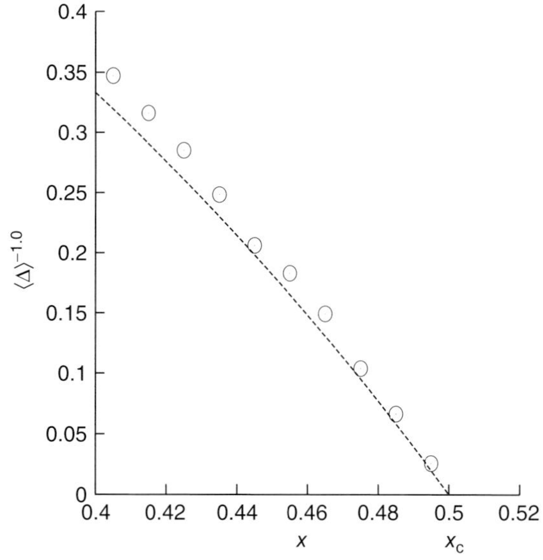

Figure 6.11 Inverse of average burst size is plotted against x for the same data set as in Figure 6.10. The solid line is the theoretical expression for the inverse of burst size versus elongation x.

$$\tau \simeq \frac{1}{4}\,(\sigma_c - \sigma)^{-\theta} \text{ with } \theta = \frac{1}{2}. \tag{6.33}$$

The susceptibility (χ) and the relaxation time (τ) follow power laws with exponent $-1/2$ with increasing external stress, and both diverge at the critical stress. If we plot χ^{-2} and τ^{-2} against external stress, we expect a linear fit near the critical point and the straight lines should cross x axis at the critical stress. We indeed found similar behavior (Figure 6.12) in simulation experiments, even for a single sample (Figure 6.13). Thus, the failure point σ_c may be estimated by extrapolating χ^{-2} and τ^{-2}.

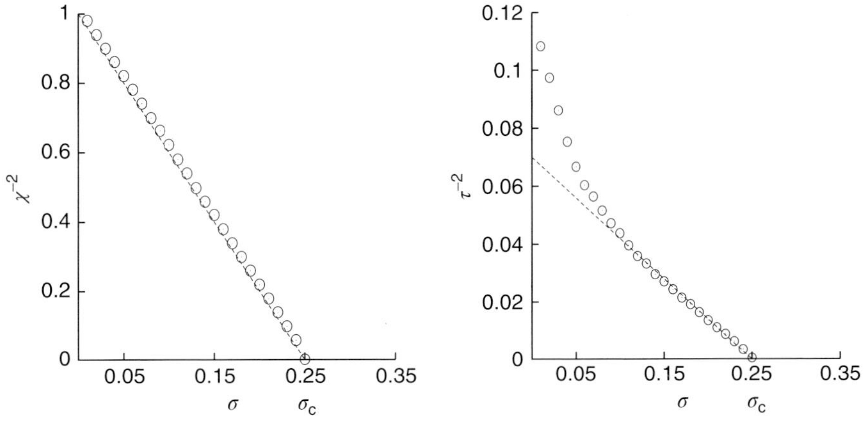

Figure 6.12 Variation of χ^{-2} and τ^{-2} with applied stress for a bundle with $N = 50\,000$ averaging over 10 000 samples. The dotted straight lines are the best linear fits near the critical point.

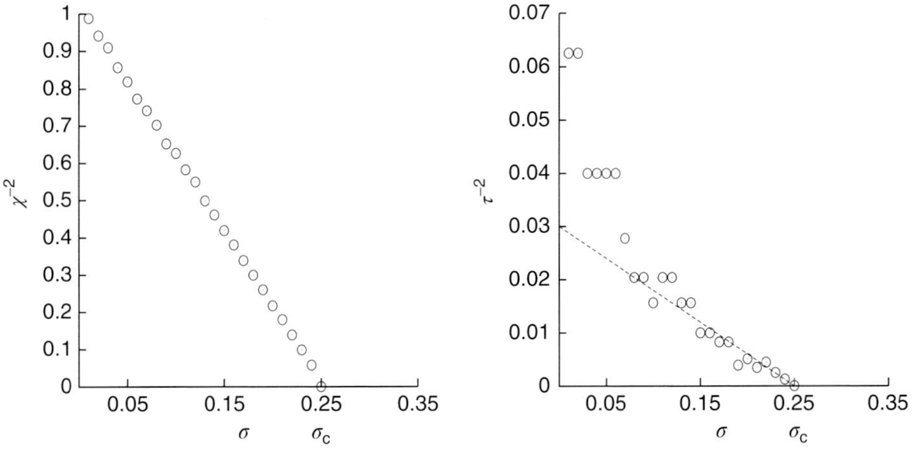

Figure 6.13 Variation of χ^{-2} and τ^{-2} with applied stress for a single bundle of 10^7 fibers with uniform distribution of fiber strengths. The straight line is the best linear fit near the critical point.

6.4
Over-Loaded Situations

What happens if the initial applied load $F = N\sigma$ is larger than the critical load of the bundle ? The stepwise failure process continues (Chapter 5) and the bundle collapses at some step t_{f}. Here the collapse is certain – but can we predict the collapse point by observing the degradation process? Once started, the failure process does not stop before complete collapse and there will be only one "burst" of size N. Here we cannot define/measure susceptibility and relaxation time due to the absence of intermediate stable states. However, we can define and measure the breaking rate and energy release rate and observe their variation with time during the failure process.

6.4.1
Breaking Rate of Loaded Fiber Bundles

Let us consider the stepwise failure process in the bundle, when the fixed applied load $F = N\sigma$ is slightly above the critical load $F_{\mathrm{c}} = N\sigma_{\mathrm{c}}$. We recall, from Chapter 5, the recursion relation (5.7)

$$n_{t+1} = 1 - P\left(\frac{\sigma}{n_t}\right) \tag{6.34}$$

in terms of the ratio $n_t = N_t/N$.

We will now demonstrate that there is a relation between the minimum of the breaking rate $R(t) = -\mathrm{d}n_t/\mathrm{d}t$ (treating t as continuous) and the moment t_{f} when the complete fiber bundle collapses.

6.4.1.1 Uniform Distribution

First we consider the simple uniform distribution, $P(x) = x$ for $0 \le x \le 1$, and assume that the load is just above the critical load: $\sigma = 1/4 + \varepsilon$, with $\varepsilon > 0$. Simulations show that the breaking rate has a minimum at some value $t_0(\varepsilon)$ and that, for varying ε, the minima all occur at a value close to $1/2$ when plotted as a function of the scaled variable t/t_{f} (Figure 6.14).

This can be shown analytically. The iteration (6.34) takes in this case the form

$$n_{t+1} = 1 - (\frac{1}{4} + \varepsilon)\frac{1}{n_t}. \tag{6.35}$$

By direct insertion, one verifies that

$$n_t = \frac{1}{2} - \sqrt{\varepsilon}\tan(At - B), \tag{6.36}$$

where

$$A = \arctan(2\sqrt{\varepsilon}) \text{ and } B = \arctan(1/2\sqrt{\varepsilon}), \tag{6.37}$$

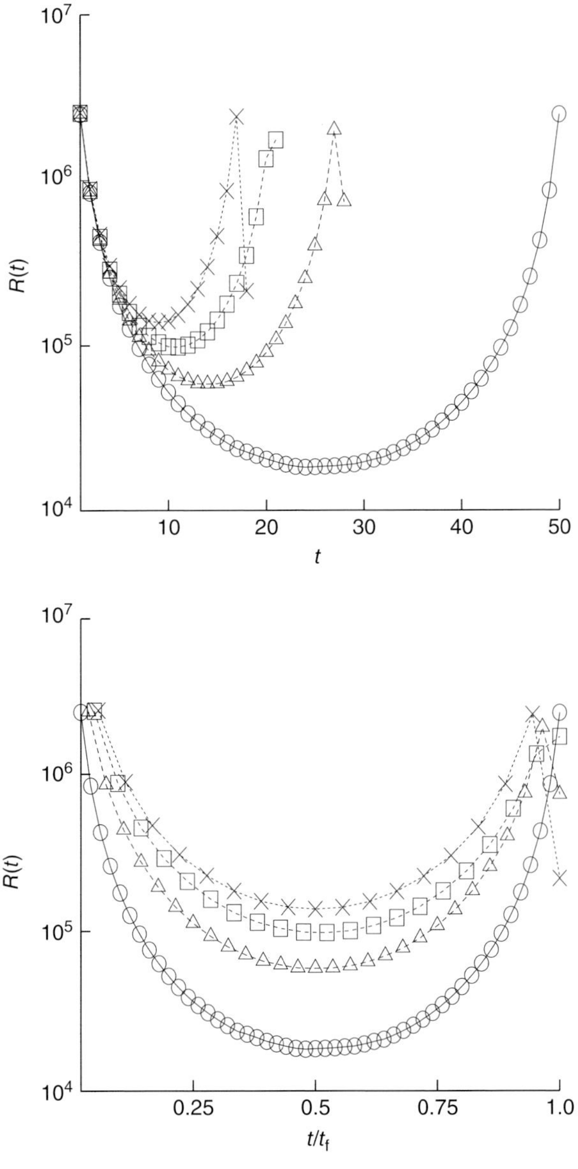

Figure 6.14 The breaking rate $R(t)$ versus step t (upper plot) versus the rescaled step variable t/t_f (lower plot) for the uniform threshold distribution for a bundle of $N = 10^7$ fibers. Different symbols are used for different excess stress levels $\sigma - \sigma_c$: 0.001 (circles), 0.003 (triangles), 0.005 (squares), and 0.007 (crosses).

is the solution to (6.35) satisfying the initial condition $n_0 = 1$. From (6.36) follows the breaking rate

$$R(t) = -\frac{\mathrm{d}n_t}{\mathrm{d}t} = \sqrt{\varepsilon}A\cos^{-2}(At - B). \tag{6.38}$$

$R(t)$ has a minimum when

$$0 = \frac{\mathrm{d}R}{\mathrm{d}t} \propto \sin(2At - 2B), \tag{6.39}$$

which corresponds to

$$t_0 = \frac{B}{A}. \tag{6.40}$$

When criticality is approached, that is, when $\varepsilon \to 0$, we have $A \to 0$, and thus $t_0 \to \infty$, as expected.

We see from Eq. (6.36) that $n_t = 0$ for

$$t_\mathrm{f} = (B + \arctan(1/2\sqrt{\varepsilon})/A = 2B/A. \tag{6.41}$$

This is an excellent approximation to the integer value at which the fiber bundle collapses completely.

Thus, with very good approximation, we have the simple connection

$$t_\mathrm{f} = 2t_0. \tag{6.42}$$

When the breaking rate starts increasing, we are halfway to complete collapse! Is it a universal feature? To test the robustness of this result, we will analyze the situation for different types of threshold distributions.

6.4.1.2 Weibull Distribution

Let us finally consider a completely different threshold distribution, a Weibull distribution of index 5, $P(x) = 1 - e^{-x^5}$. Simulations reveal that the breaking rate has a similar behavior as in the two cases considered (Figure 6.15).

This case is more complicated, but the analytical ground work has already been done in Chapter 5. From Eqs. (5.84) and (5.86), we can see that for small ε the iteration is of the form

$$n_t = n_\mathrm{c} - b\sqrt{\varepsilon/C}\,\tan(t\sqrt{C\varepsilon} - c). \tag{6.43}$$

Here $n_\mathrm{c} = e^{-1/5}$, $C = 5(5e)^{1/5}/2$, $b = 5^{1/5}$, and the constant c is determined by the initial condition $n_0 = 1$:

$$c = \arctan[(1 - n_\mathrm{c})b^{-1}\sqrt{C/\varepsilon}]. \tag{6.44}$$

From (6.43), the breaking rate equals

$$R(t) = -\frac{\mathrm{d}n_t}{\mathrm{d}t} \propto \cos^{-2}(t\sqrt{C\varepsilon} - c). \tag{6.45}$$

The breaking rate is a minimum when the cosine takes its maximum value 1. This is the case when

$$t_0 = \frac{c}{\sqrt{C\varepsilon}} = (C\varepsilon)^{-1/2}\arctan[(1 - n_\mathrm{c})b^{-1}\sqrt{C/\varepsilon}]. \tag{6.46}$$

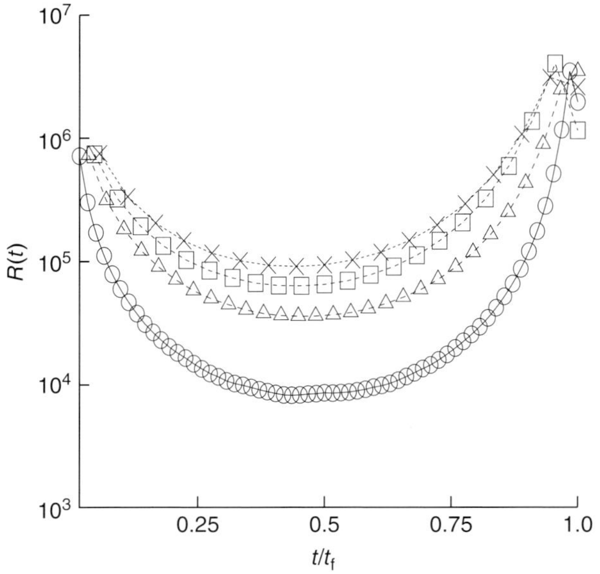

Figure 6.15 Simulation result for the breaking rate $R(t)$ versus the rescaled step variable $t/t_{\rm f}$ for a bundle of $N = 10^7$ fibers having a Weibull threshold distribution of index 5. Different symbols are used for different excess stress levels $\sigma - \sigma_{\rm c}$: 0.001 (circles), 0.003 (triangles), 0.005 (squares), and 0.007 (crosses).

The inverse tangent is close to $\pi/2$ when ε is very small. Hence, for small overloads, we have in excellent approximation

$$t_0 = \frac{\pi}{2\sqrt{C\varepsilon}}. \tag{6.47}$$

The collapse point $t_{\rm f}$ is already evaluated in Chapter 5, with the result (5.94), which can be written in terms of C as

$$t_{\rm f} \simeq \frac{\pi}{\sqrt{C\varepsilon}} \tag{6.48}$$

for small ε.

Comparison between (6.48) and (6.47) gives

$$t_{\rm f} \simeq 2t_0, \tag{6.49}$$

as for the previous threshold distribution.

6.4.1.3 Large Overload Situation

We have shown that the complete collapse of fiber bundles occurs at $t_{\rm f} = 2t_0$, where t_0 denotes the number of steps of the breaking process at which the fiber breaking rate has a minimum. The results are derived for very small overloads ε.

For larger overloads, the ratio t_0/t_f will not be exactly 0.5 but nevertheless of the order of 0.5.

Another interesting observation is that at $t = t_0$ the number of unbroken fibers in the bundle $n(t_0)$ attains the critical value n_c. This can be derived analytically by putting the value of t_0 into the expressions of n_t for, respectively, the uniform and the Weibull distribution.

6.4.2 Energy Emission Bursts

So far, we have observed that fiber bundles under external load produce bursts (avalanches) of different magnitudes, where a burst consists of simultaneous rupture of several elements. At each failure, the sudden internal stress redistribution in the material is accompanied by a rapid release of mechanical energy. Therefore, with each *burst* there will be a corresponding *energy emission* burst. A useful experimental technique to monitor such energy bursts is to measure the acoustic emissions (AEs), the elastically radiated waves produced in the bursts [81].

We now show that the catastrophic collapse point of an overloaded bundle can be predicted by monitoring the energy emission rate [82].

We recall that the failure process in a loaded fiber bundle model proceeds via the recursion formula (5.7)

$$n_{t+1} = 1 - P\left(\frac{\sigma}{n_t}\right), \tag{6.50}$$

in terms of n_t, the fraction of undamaged fibers at step t. At each iteration step, a certain amount of elastic energy is released and we consider now the energy emission process.

For a burst in which the number of intact fibers is reduced from N_{t-1} to N_t, all fibers with thresholds x between the values $F/N_{t-1} = \sigma/n_{t-1}$ and $F/N_t = \sigma/n_t$ break. Since there are $N p(x)\,\mathrm{d}x$ fibers with thresholds in $(x, x + \mathrm{d}x)$, the energy emitted in this burst is given by

$$E_t = \int_{\sigma/n_{t-1}}^{\sigma/n_t} \frac{1}{2}x^2\, N p(x)\,\mathrm{d}x. \tag{6.51}$$

We consider external stresses that are slightly above the critical value,

$$\sigma = \sigma_c + \varepsilon, \tag{6.52}$$

where ε is small and positive. Simulations for a uniform threshold distribution, $p(x) = 1$ for $0 \le x \le 1$, show that the energy emission has a minimum at some value $t_E(\varepsilon)$ and that, for varying loads, the minima all occur at value close to $1/2$ when plotted as a function of the scaled variable t/t_f (Figure 6.16). Here, t_f is the number of iterations corresponding to complete bundle failure. We will show this analytically and, in addition, demonstrate that the result is not limited to the uniform distribution.

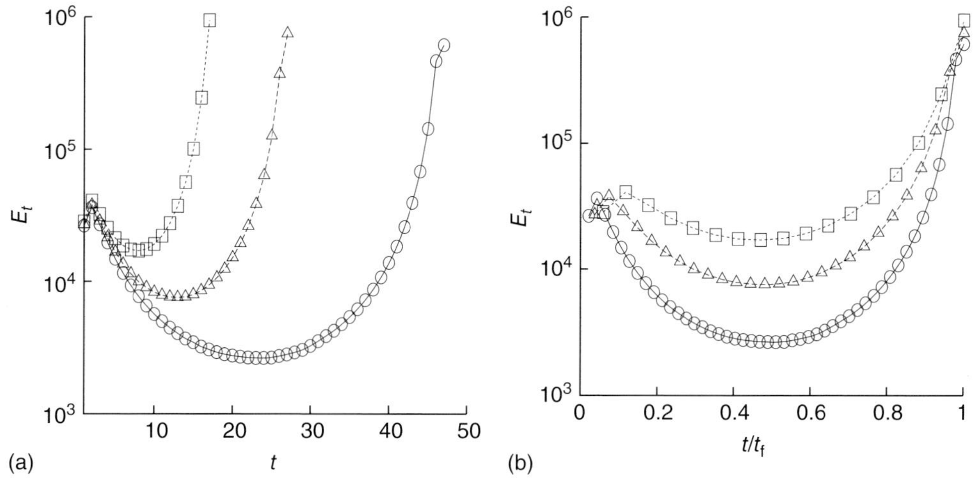

Figure 6.16 Simulation results for the energy emission E_t versus step t (a) and versus the rescaled step variable t/t_f (b) for the uniform threshold distribution for a bundle of $N = 10^7$ fibers. Different symbols are used for different excess stress levels $\sigma - \sigma_c$: 0.001 (circles), 0.003 (triangles), 0.007 (squares).

Figure 6.16 shows that for large N and near the minimum, the energy emission E_t appears to be an almost continuous function of t. The minimum is therefore located where the derivative of E_t with respect to t vanishes. From (6.51), we obtain

$$\frac{\mathrm{d}E_t}{\mathrm{d}t} = \frac{1}{2}N[-\sigma^2 n_t^{-4} p(\sigma/n_t)\, \dot{n}_t + \sigma^2 n_{t-1}^{-4}\, p(\sigma/n_{t-1})\, \dot{n}_{t-1}], \tag{6.53}$$

where the dot denotes differentiation with respect to time.

To eliminate quantities at time step $t-1$, we use the connection (6.50), which gives

$$\dot{n}_t = \dot{n}_{t-1}\, p(\sigma/n_{t-1})\, \sigma n_{t-1}^{-2}. \tag{6.54}$$

Insertion into (6.53) yields

$$\frac{\mathrm{d}E}{\mathrm{d}t} = \dot{n}_t\, \frac{N}{2\sigma}\left[-\frac{\sigma^3 p\left(\sigma/n_t\right)}{n_t^4} + \frac{\sigma^2}{n_{t-1}^2}\right]. \tag{6.55}$$

We also need n_{t-1}. Equation (6.50) gives

$$\frac{\sigma}{n_{t-1}} = P^{-1}(1 - n_t), \tag{6.56}$$

where P^{-1} denotes the inverse function to P. In conclusion,

$$\frac{\mathrm{d}E}{\mathrm{d}t} = \dot{n}_t\, \frac{N}{2\sigma}\left[-\frac{\sigma^3 p\left(\sigma/n_t\right)}{n_t^4} + (P^{-1}(1 - n_t))^2\right]. \tag{6.57}$$

As $\dot{n}_t$ is always negative, the minimum of energy emission occurs when the relative number of intact fibers satisfies

$$P^{-1}(1 - n_t) = \sqrt{\sigma^3\, p\left(\frac{\sigma}{n_t}\right)\, n_t^{-4}}, \tag{6.58}$$

that is,

$$1 - n_t = P\left(\sqrt{\sigma^3\, p\left(\frac{\sigma}{n_t}\right)\, n_t^{-4}}\right). \tag{6.59}$$

Now we turn to specific cases. We start with the simplest case, the uniform distribution, on which the simulations in Figure 6.16 were based.

6.4.2.1 Uniform Distribution

For the uniform threshold distribution, $P(x) = x$ for $0 \leq x \leq 1$, the condition (6.59) takes the form

$$1 - n_t = \sqrt{\sigma^3\, n_t^{-4}}, \tag{6.60}$$

or

$$(1 - n_t)n_t^2 = \sigma^{3/2}. \tag{6.61}$$

For this case, the maximum of the force is $F = N/4$, corresponding to the critical stress $\sigma_c = 1/4$. For a small excess stress $\varepsilon = \sigma - \sigma_c = \sigma - 1/4$, Eq. (6.61) takes the form

$$(1 - n_t)n_t^2 = \left(\frac{1}{4} + \varepsilon\right)^{3/2} = \frac{1}{8}(1 + 6\varepsilon + \mathcal{O}(\varepsilon^2)). \tag{6.62}$$

For the lowest order, $n_t = 1/2$, and one shows easily that

$$n_t = \frac{1}{2} + 3\varepsilon \tag{6.63}$$

satisfies (6.62) to the first order in ε.

To find t_E, the value of t corresponding to (6.63) we use the previously derived solution of the iteration (6.50)

$$n_t = \frac{1}{2} - \sqrt{\varepsilon}\, \tan(At - B), \tag{6.64}$$

where $A = \arctan(2\sqrt{\varepsilon}) \simeq 2\sqrt{\varepsilon}$ and $B = \arctan(1/2\sqrt{\varepsilon})$.

We see that (6.63) requires

$$\tan(At_E - B) = -3\sqrt{\varepsilon}, \tag{6.65}$$

that is,

$$t_E - B/A = -3\sqrt{\varepsilon}/A. \tag{6.66}$$

To the lowest order, we therefore obtain

$$t_E = B/A - \frac{3}{2}. \tag{6.67}$$

Moreover, we see from (6.64) that at for

$$t = t_{\rm f} = 2B/A \tag{6.68}$$

we have

$$n_{t_{\rm f}} = \frac{1}{2} - \sqrt{\varepsilon}\ \tan(B) = 0, \tag{6.69}$$

signifying complete collapse of the fiber bundle. By comparison between (6.67) and (6.68), we obtain

$$t_E = \frac{1}{2}\ t_{\rm f} - \frac{3}{2}. \tag{6.70}$$

Thus, the minimum energy emission occurs almost halfway to complete bundle collapse, in agreement with the simulations shown in Figure 6.16.

6.4.2.2 Weibull Distribution

To illustrate the generality of the connection between the energy emission minimum and the bundle collapse, we now turn to a completely different threshold distribution, namely, a Weibull distribution of index 5,

$$P(x) = 1 - e^{-x^5}. \tag{6.71}$$

The critical stress for this distribution is $\sigma_{\rm c} = (5e)^{-1/5} = 0.5933994$. Simulations reveal a qualitatively similar behavior as for the uniform distribution (Figure 6.17).

The condition (6.59) for minimal energy emission takes in this case the form

$$-\ln\ n_t - 5^{5/2}\sigma^{35/2}\ n_t^{-20}\ \exp\ (-5\sigma^5/2n_t^5) = 0. \tag{6.72}$$

It is straightforward to verify that for $\sigma = \sigma_{\rm c} = (5e)^{-1/5}$ (6.72) has the solution $n_t = e^{-1/5}$. We are interested in a slightly overloaded bundle,

$$\sigma = \sigma_{\rm c} + \varepsilon = (5e)^{-1/5} + \varepsilon, \tag{6.73}$$

with ε small and positive, and we seek the corresponding value of n_t. Putting (6.73) and $n_t = e^{-1/5} + \delta$ into the condition (6.72), and expanding in ε and δ, we obtain to the first order in small quantities

$$e^{1/5}\ \delta = 3(5e)^{1/5}\ \varepsilon - \frac{7}{2}\ e^{1/5}\ \delta. \tag{6.74}$$

Thus, $\delta = \frac{2}{3}\ 5^{1/5}\ \varepsilon$. Hence,

$$n_t = e^{-1/5} + \frac{2}{3}\ 5^{1/5}\ \varepsilon \tag{6.75}$$

is the relative number of undamaged fibers when the energy emission is minimal.

To find t_E, the value of t corresponding to (6.75), we take advantage of the ground work already done in Chapter 5 for the Weibull distribution. Equations (5.84) and (5.86) show that for small ε the iteration is of the form

$$n_t = e^{-1/5} - 5^{1/5}\sqrt{\varepsilon/C}\ \tan(t\sqrt{\varepsilon C} - c), \tag{6.76}$$

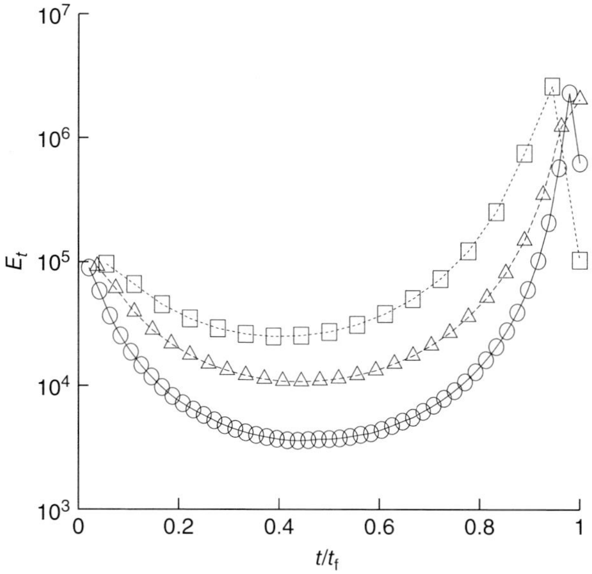

Figure 6.17 Simulation results for the energy emission E_t versus the scaled step variable t/t_f for a bundle of $N = 10^7$ fibers obeying the Weibull threshold distribution (6.71). Different symbols are used for different excess stress levels $\sigma - \sigma_c$: 0.001 (circles), 0.003 (triangles), 0.007 (squares).

with $C = 5(5e)^{1/5}/2$ and the constant $c = \arctan[(1 - e^{-1/5})\, 5^{-1/5} \sqrt{C/\varepsilon}]$ ensures that the initial condition $n_0 = 1$ is satisfied.

Comparison between (6.75) and (6.76) gives

$$\tan(t_E\sqrt{\varepsilon C} - c) = -\frac{2}{3}\sqrt{\varepsilon C}. \tag{6.77}$$

To the dominating order in ε, then,

$$t_E\sqrt{\varepsilon C} - c = -\arctan\left(\frac{2}{3}\sqrt{\varepsilon C}\right) \simeq -\frac{2}{3}\sqrt{\varepsilon C}. \tag{6.78}$$

For small ε, the constant c is very close to $\pi/2$, so that in good approximation we have

$$t_E = \frac{\pi}{2\sqrt{\varepsilon C}} - \frac{2}{3}. \tag{6.79}$$

The collapse time t_f was also evaluated in Chapter 5, with the result (5.94),

$$t_f = \frac{\pi}{\sqrt{\varepsilon C}}. \tag{6.80}$$

Consequently, we have

$$t_E = \frac{1}{2}\, t_f - \frac{2}{3}. \tag{6.81}$$

In excellent approximation, the minimum of energy emissions occurs halfway to the final collapse also for the Weibull distribution. There is every reason to believe that this feature is general. Note that since t_E and t_f are large numbers, the constants 3/2 and 2/3 in (6.70) and (6.81) are of no significance.

We emphasize that our concern is small values of the stress in excess of the critical stress. For large stress values, the system breaks down quickly. Considerably more interesting are situations in which the system is apparently labile, so that the breakdown takes a long time. Should the energy emission of such a system be observed to possess a minimum at time t_E, then this foreshadows a complete breakdown at a later time t_f. In the following, we show for our model systems that for excess stresses less than 9–10 per cent of the critical stress, such an emission minimum will be present.

6.4.2.3 Large Overload Situation

So far, we have considered energy emissions from a slightly overloaded bundle, for which the breakdown process proceeds slowly. Here we give a brief description of where the energy emission minimum is located for larger loads. As a specific example, we consider the model with a uniform threshold distribution.

For a given stress σ, the relative number of unbroken fibers at the emission minimum is given by Eq. (6.61),

$$(1 - n_t)n_t^2 = \sigma^{3/2}. \tag{6.82}$$

Since the left-hand side of (6.82) does not exceed 4/27, the maximum stress when an emission minimum is present equals

$$\sigma_\mathrm{m} = 2^{4/3}/9 = 0.279982. \tag{6.83}$$

For a given stress, $\sigma_\mathrm{c} \leq \sigma \leq \sigma_\mathrm{m}$, Eq. (6.82) determines n_t, and the corresponding value of t_E/t_f can be calculated using

$$n_t = \frac{1}{2} + \sqrt{\varepsilon}\tan[B(1 - 2t_E/t_\mathrm{f})], \tag{6.84}$$

with $B = \arctan(1/2\sqrt{\varepsilon})$. Figure 6.18 shows the result. The position of the emission minimum, relative to t_f, decreases slowly with increasing stress.

In this derivation, the integer t is considered a continuous variable. This treatment is satisfactory for slightly overloaded bundles, but less so when t_E is smaller. To find the smallest stress that produces a minimum, we consider the first few energy emissions. Equation (6.51) for the uniform threshold distribution,

$$E_t = \frac{1}{6}N\sigma^3(n_t^{-3} - n_{t-1}^{-3}), \tag{6.85}$$

and the iteration (6.50), $n_0 = 1, n_1 = 1 - \sigma, n_2 = (1 - 2\sigma)/(1 - \sigma)$, etc., gives

$$E_0 = \sigma^3 \tag{6.86}$$

$$E_1 = \left(\frac{\sigma}{1 - \sigma}\right)^3 - \sigma^3 \tag{6.87}$$

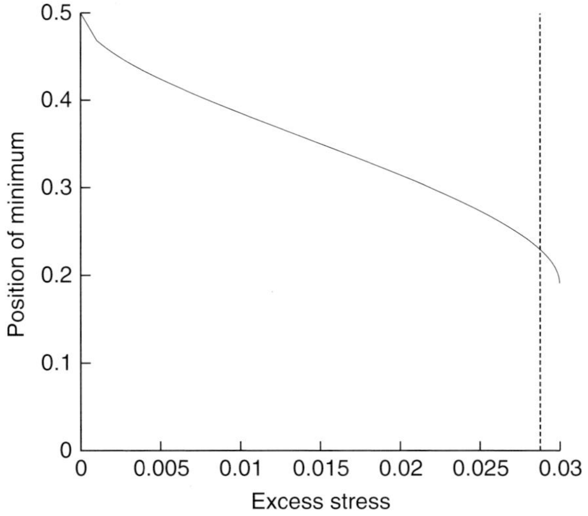

Figure 6.18 Position of the emission minimum in terms of t_E/t_f versus excess stress for the uniform fiber strength distribution, computed using the continuous time approximation. The dashed line marks the maximum stress with an emission minimum.

$$E_2 = \left(\frac{\sigma(1-\sigma)}{1-2\sigma}\right)^3 - \left(\frac{\sigma}{1-\sigma}\right)^3 \tag{6.88}$$

$$E_3 = \left(\frac{\sigma(1-2\sigma)}{1-3\sigma+\sigma^2}\right)^3 - \left(\frac{\sigma(1-\sigma)}{1-2\sigma}\right)^3, \tag{6.89}$$

where the common factor $N/6$ is omitted. For all subcritical stresses $E_1 > E_0$, but a local minimum may occur already at $t = 2$. It is easy to show that

$$E_1 = E_2 \quad \text{for } \sigma = \sigma_1 = 0.278764 \tag{6.90}$$

$$E_2 = E_3 \quad \text{for } \sigma = \sigma_2 = 0.273054. \tag{6.91}$$

For $\sigma \geq \sigma_1$ the energy emission bursts increase monotonically in size. For $\sigma_2 < \sigma < \sigma_1$, we have $E_2 < E_1$ and also $E_2 < E_3$. Hence, there is a local emission minimum at $t_E = 2$ in this range.

Thus, a local emission minimum is no longer present when the stress value exceeds σ_1. For the limiting value $\sigma = \sigma_1$, the bundle collapse occurs at $t_f = 8$, signified by the first nonpositive value of n_t. This gives $t_E/t_f = 0.25$ at $\sigma = \sigma_1$.

This limiting value is a little lower than the value σ_m, the end point of the graph in Figure 6.18, obtained by the continuous procedure. However, at the maximum stress σ_1 that produces a local emission minimum, the value of t_E/t_f equals 0.230, not far from the exact value 0.25. We conclude that the graph in Figure 6.18 is very precise for small supercritical stresses and fairly accurate up to the stress σ_1.

Let us also consider the Weibull distribution of index 5. The energy emission has a minimum when Eq. (6.72) has a solution. For a given value of σ, the left-hand side of Eq. (6.72), considered as a function of n_t, exhibits a maximum. The maximum must be nonnegative for Eq. (6.72) to have a solution, and we find that this is the case for supercritical stresses below $\sigma_1 = 0.6457$. This means that the emission minimum exists for an excess stress less than

$$\sigma_1 - \sigma_c = 0.0523. \tag{6.92}$$

6.4.3
Energy Burst Pattern

As mentioned before, energy releases in the form of AEs occur during fracturing process under external load. Such AE signals can easily be measured by placing acoustic transducers around the sample. In Figure 6.19, we present a plot showing records of AE events with time during a rock fracturing test where the external load is increased slowly until the collapse of the sample. It is not difficult to study the statistics of such AE events. Can this AE statistics tell us about the load level on the sample?

If we record the energy E_t at each step of load redistribution in a fiber bundle, different pattern will be seen depending on the stress level, being critical, supercritical, or subcritical (Figure 6.20). If we record such energy emission bursts separately for below-critical and over-critical levels, the corresponding distributions exhibit marked differences (Figures 6.21 and 6.22). For stresses below the

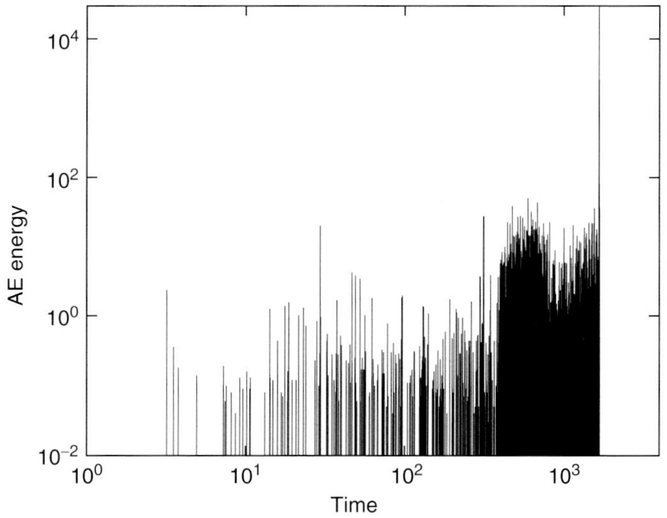

Figure 6.19 Energies of AE signals recorded during a rock fracturing test on Castlegate Sandstone sample at SINTEF lab.

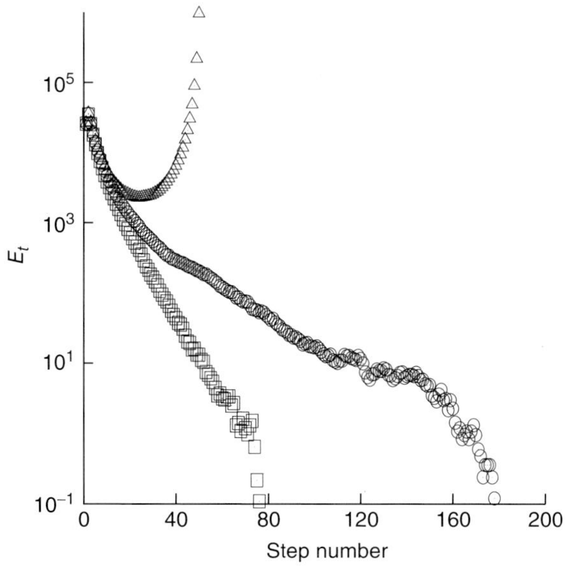

Figure 6.20 Energy emission E_t versus step number t in the load redistribution process in a bundle with $N = 10^7$ fibers having uniform strength distribution within the unit interval. Different symbols indicate different stress levels: critical stress σ_c (circles); $\sigma = \sigma_c + 0.001$ (triangles), and $\sigma = \sigma_c - 0.001$ (squares).

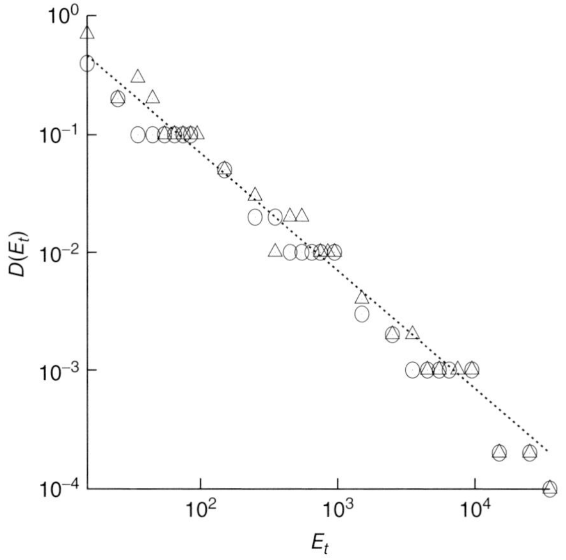

Figure 6.21 Distribution of energy emissions E_t in the same fiber bundle as in Figure 6.20 for sub-critical stresses: $\sigma = \sigma_c - 0.001$ (circles), $\sigma = \sigma_c - 0.002$ (triangles). The straight line has a slope -1.0.

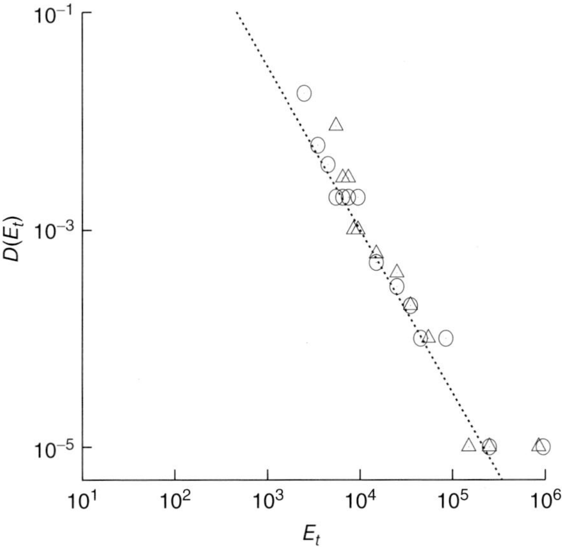

Figure 6.22 Distribution of energy emissions E_t in the same fiber bundle as in Figure 6.20 for stresses above the critical value: $\sigma = \sigma_c + 0.001$ (circles), $\sigma = \sigma_c + 0.002$ (triangles). The straight line has a slope −1.5.

critical level, there are many small energy bursts, which are absent in the latter case. The exponents of the energy burst distributions are different for these two situations, equal to −1 and −1.5, respectively.

7
Fiber Bundle Model in Material Science

In this chapter, we demonstrate, through some representative illustrations, how the fiber bundle model can be used as a tool for studying important phenomena in materials, particularly in fibrous materials.

Through plastic deformations, a material may become stronger. A simple example is that bending a paper clip several times makes the metal stronger at the bending point. This phenomena is called *work hardening* or strain hardening. For a fiber bundle to exhibit work hardening, the fibers cannot have merely two states, intact or failed. There must also be states where the fibers are not completely broken, but merely have reduced strength. We shall see in Section 7.1 that when such strength-reducing events can occur, several types of bundle strength behavior may be modeled.

Creep is the slow deformation of a material under the influence of a constant force. Creep will end in complete material failure, even if the material throughout is subjected to subcritical stresses. The time before creep rupture – the lifetime of the material – decreases with increasing load. The underlying microscopic mechanisms of creep rupture are complex, depending on several characteristics of the specific properties of the material. In Section 7.2, we describe general characteristics of creep failure by modeling how the fibers are continuously weakened through damage accumulation.

In Section 7.3, we treat *viscoelastic creep* in fiber bundles with fibers that are both elastic and viscous. In this case, the stress will have a critical value, above which the fate of the bundle is failure, and near this stress value, critical behavior is seen.

Fatigue is the progressive damage that occurs when a material is subjected to cyclic loading or, more generally, to repeated loading and unloading. The material is throughout exposed to subcritical stresses; hence, fatigue rupture is a time-dependent phenomenon. We report in Section 7.4 on an interesting experimental study of fatigue in asphalt, modeled as a fiber bundle.

Since in creep the strain accumulates as a result of long-time stress, creep is a time-dependent phenomenon. More general time-dependent events are caused by the *influence of noise* on the bundle dynamics. Thermal noise is a well-defined source, but one can easily imagine other types of disturbances. The influence of temperature on a fiber bundle is, in Section 7.5, considered by assuming that the

The Fiber Bundle Model: Modeling Failure in Materials, First Edition.
Alex Hansen, Per C. Hemmer and Srutarshi Pradhan.

Gaussian distributed thermal fluctuations give an extra contribution to the stress on a fiber, in addition to the one caused by the external force.

In Section 7.6, we also consider the influence of noise in a more empirical way, without specifying the precise mechanisms. When the noise level is low, the failure process may proceed in two different modes: a slow mode with waiting times between each noise-induced failure or a rapid continuous fracturing mode induced by load transfer from the previous failure. In the stress–noise plane, one may define a "phase boundary" that separates these two types of rupture phases.

In a *fiber-reinforced composite*, fibers are embedded in a matrix to form the composite. During tensile loading, the main part of the load is carried by the fibers and the strength of the composite is governed to a large extent by the strength of the fibers. The matrix material is chosen so that its yield threshold is lower than that of the fibers that are embedded in it. Common materials used for the fibers are aluminum, aluminum oxide, aluminum silica, asbestos, beryllium, beryllium carbide, beryllium oxide, carbon (graphite), glass, molybdenum, polyamide (aromatic polyamide, aramid), Kevlar, polyester, quartz (fused silica), steel, tantalum, titanium, and tungsten or tungsten monocarbide. Most matrix materials are resins as a result of their wide variation in properties and relatively low cost. Common resin materials are epoxy, phenolic, polyester, polyurethane, and vinyl ester. When the composite is to be used under adverse conditions such as high temperature, metallic matrix materials such as aluminum, copper, lead, magnesium, nickel, silver, or titanium or nonmetallic matrix materials such as ceramics may be used. When the matrix material is brittle, cracks open up in the matrix perpendicular to the fiber direction at roughly equal spacing. In metallic matrix materials, plasticity sets in at sufficient load. Finally, in polymer matrix composites, the matrix typically responds linearly, but still the fibers carry most of the load due to the large compliance of the matrix. When a fiber fails, the forces it carried are redistributed among the surviving fibers and the matrix. If the matrix–fiber interface is weak compared to the strengths of the fibers and the matrix, fractures develop along the fibers. When the matrix is brittle, the fibers bridging the developing crack in the matrix will, besides binding the crack together, lead to stress alleviation at the matrix crack front. This leads to the energy necessary to propagate a crack increases further with the length of the crack. When the bridging fibers fail, they typically do so through debonding at the fiber–matrix interface. This is followed by pull-out. A good general reference for further reading is Mallick's book [11].

7.1 Repeated Damage and Work Hardening

Work hardening (also called strain hardening) is the process of making a material stronger through plastic deformation. When there are so many dislocations in a metal that they impede each other, work hardening sets in.

In order for model work hardening in fiber bundles, one must consider more general types of fibers than have been considered until now. In the previous chapters, we assumed that each fiber is in one of two states, either intact or failed. Here it will be realistic to consider a wider class of fibers, in which each fiber has an internal structure that at a certain threshold may be damaged without failing completely. As a result, the fiber will continue to contribute to the bundle strength, but with an elastic constant that is reduced by some factor a [84]. To make the model precise, it is assumed here that when the load on a fiber i is reduced from σ_i to $a\sigma_i$, the rest of the load is distributed among the other fibers. Note that this form of load sharing is not *identical* to the usual equal-load-sharing rule, as now stiffer fibers receive more load. While this method of load distribution is simple, it is not the only possibility. Several local-load-sharing mechanisms are also possible in this case.

The model may be generalized to include fibers that may be damaged any number of times.

7.1.1 The Load Curve

For simplicity, we assume that the bundle size N is so large that one can safely neglect fluctuations. Thus, we concentrate on the macroscopic (or average) force law. The simplest case is when the fibers can be damaged just once. Let $P(x)$ be the cumulative distribution for the damage thresholds. Thus, the average fraction of damaged fibers at elongation x is $P(x)$. Then the average force per fiber (the load curve) equals

$$\sigma_1(x) = x[1 - P(x)] + a\, x\, P(x), \tag{7.1}$$

where we as usual have set the elastic constant equal to unity. The first term is the contribution from the intact fibers, and the last term represents the average load carried by the damaged fibers.

As a concrete example, let $P(x)$ be a Weibull distribution of index k,

$$P(x) = 1 - e^{-x^k}. \tag{7.2}$$

In Figure 7.1, two cases are shown, one with $a = 0.5$ for Weibull index 2 and the other with $a = 0.4$ for Weibull index 6. Note that for large strain, the load curves are linear rather than drop to zero force. This represents work hardening.

The so-called micromechanical model of fiber-reinforced ceramic matrix composites, in which $a = 0.5$ is taken as a typical value, has been much studied in the literature [85]. In this model, the physical origin of the load-bearing capacity is that near the damaged fiber, the fiber–matrix interface debonds and the stress builds up again in the damaged fiber through the sliding fiber–matrix interface.

7.1.2 Repeated Damages

In principle, a fiber may be damaged more than once [84, 86]. If a fiber may be damaged twice, and each time the same strength reduction factor a applies, the

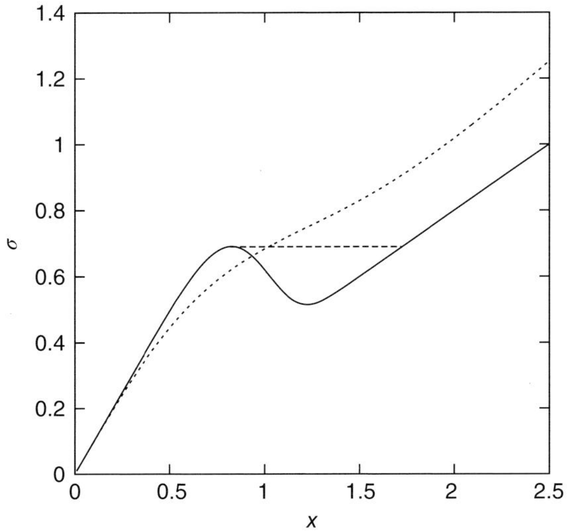

Figure 7.1 Load curves $\sigma_1(x)$ for Weibull fibers that can be damaged just once. The dotted curve is for the case of fibers with Weibull index 2, when a damaged fiber has its strength reduced by 50%. The fully drawn curve corresponds to Weibull index 6, and a strength reduction factor $a = 0.40$. Note that in a force-controlled experiment, the minimum is not realized, the system will follow the least monotonic function of $\sigma(x)$, as indicated.

load curve takes the form

$$\sigma_2(x) = x[1 - P(x)] + ax\ [P(x) - P(ax)] + a^2xP(ax). \tag{7.3}$$

Here the last term represents fibers twice damaged, and $P(x) - P(ax)$ in the second term corresponds to the fraction of fibers damaged merely once. We have assumed that the damage threshold remains the same for the second failure (the so-called quenched situation). Another possibility (the annealed situation), not discussed here, would be that after each damage a new threshold is picked from the same statistical distribution.

If m damages is allowed for each fiber, the load curve generalizes to

$$\sigma_m = x\sum_{n=0}^{m} a^n P_n(x), \tag{7.4}$$

where $P_k(x)$ is the probability that at strain x an arbitrary fiber is damaged k times. In particular, $P_0(x) = 1 - P(x)$ is the probability of finding the fiber intact. Moreover,

$$\begin{aligned} P_n(x) &= P(a^{n-1}x) - P(a^n x) && \text{for } 1 \leq n \leq m-1 \\ P_m(x) &= P(a^{m-1}x) && \text{for } n = m. \end{aligned} \tag{7.5}$$

Equation (7.3) is the special case $m = 2$. Note that the probabilities satisfy the sum rule

$$\sum_{n=0}^{m} P_n(x) = 1, \tag{7.6}$$

as expected.

When (7.5) is inserted into (7.4), we find

$$\begin{aligned}\sigma_m(x) = x \{ 1 - P(x) + a[P(x) - P(ax)] + a^2[P(ax) - P(a^2x)] + \\ \ldots + a^{m-1}[P(a^{m-2}) - P(a^{m-1}x)] + a^m P(a^{m-1}x) \}\end{aligned} \tag{7.7}$$

$$\begin{aligned}= x(1-a) \{ 1 - P(x) + a[1 - P(ax)] + a^2[1 - P(a^2x)] + \\ \ldots + a^{m-1}[1 - P(a^{m-1}x)] \} + xa^m.\end{aligned} \tag{7.8}$$

When $a = 0$, we have the usual load curve $\sigma_0 = x[1 - P(x)]$, as expected. In general, the load curve is linear for both small and large x. As $P(0) = 0$, we have for small x that $\sigma_m(x) \simeq x$ and for large x $\sigma_m(x) \simeq a^m x$. Thus, the slope a^m of the linear asymptotic force decreases with increasing m. In Figure 7.2, Weibull load curves with different numbers m of consecutive damages are shown. For finite m, the fibers have for large strain a residual strength of $a^m x$, that is, a finite Young modulus. This represents hardening of the material. With increasing m, the hardening part is preceded by a quasiplastic plateau of increasing length. Similar behavior has been seen in brittle matrix composites, in which repeated cracking of the matrix is responsible for the quasiplastic plateau, and the asymptotic linear part is due to the remaining elastic fibers that remain intact after matrix cracking [87].

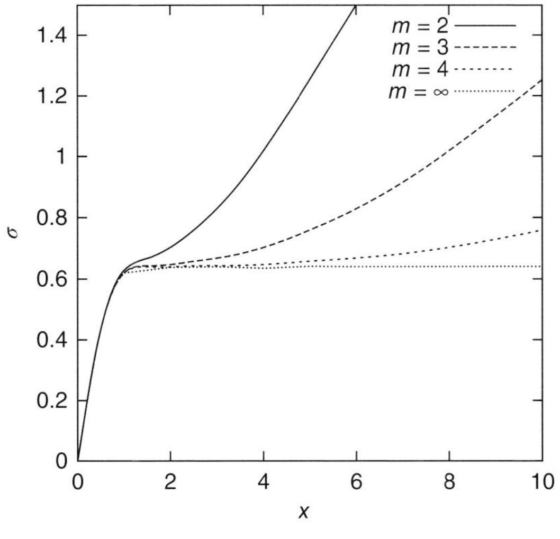

Figure 7.2 Load curves $\sigma_m(x)$ for a large bundle with thresholds following a Weibull distribution of index 2, with a strength reduction factor $a = 0.50$, and for different values of the maximum number m of damages.

We see in particular that in the limit $m \to \infty$, the force apparently becomes essentially constant beyond a certain strain, that is, the bundle behaves in this case completely plastic. Let us show how this follows from (7.8). In the limit $m \to \infty$, the last term $a^m x$ vanishes. Thus,

$$\sigma_\infty(x) = x(1-a)\sum_{n=0}^{\infty} a^n[1 - P(a^n x)]. \tag{7.9}$$

As

$$\sigma_\infty(ax) = x(1-a)\sum_{n=0}^{\infty} a^{n+1}[1 - P(a^{n+1} x)], \tag{7.10}$$

we have

$$\sigma_\infty(ax) = \sigma_\infty(x) - (1-a)x[1 - P(x)] = \sigma_\infty(x) - (1-a)\sigma_0(x). \tag{7.11}$$

As the load curve for the damage-free bundle, $\sigma_0(x)$, equals or approaches 0 for large x, we obtain

$$\sigma_\infty(ax) = \sigma_\infty(x) \quad \text{for large } x, \tag{7.12}$$

consistent with a plateau behavior.

For Weibull distribution of index 2 and with $a = 0.5$, Eq. (7.9) produces the asymptotic plateau value in Figures 7.2 and 7.3 to be approximately $\sigma_\infty = 0.6391$.

7.1.3 Damages Ending in Complete Bundle Failure

A different scenario is that a fiber, after a maximum number m of damages, experiences complete failure rather than work hardening. Then the residual stiffness of the fibers is set to zero after m failures. In other words, the last term P_m in (7.4) or (7.7) is not present. In the simplest case, with at most one nonfatal failure allowed, the load curve (7.3) takes the new form

$$\sigma_1^*(x) = x[1 - P(x)] + ax[P(x) - P(ax)]. \tag{7.13}$$

More generally,

$$\sigma_m^*(x) = \sigma_{m+1}(x) - xa^{m+1}P(a^m x), \tag{7.14}$$

where the functions $\sigma_m(x)$ are given by (7.7).

In Figure 7.3, we have exhibited several load curves $\sigma_m^*(x)$ for a bundle with Weibull fibers of index 2 and with a strength reduction factor $a = 0.5$. As one expects, the load curves extend now to higher and higher strain with increasing m. In a strain-controlled experiment, the complete load curves are followed, while in the force-controlled situation, the bundle collapses when the critical point (the maximum) is reached.

Note that when $m \to \infty$, the load curves approach a horizontal plateau, the same plateau as in the work-hardening case. In the previous case, the plateau was approached from above, in the present case, however, it is approached from below.

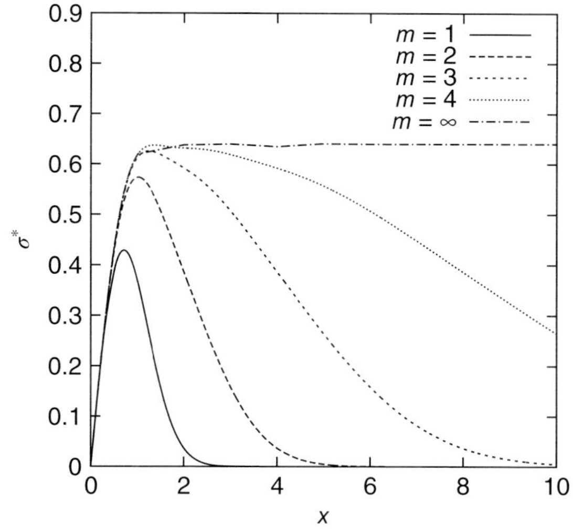

Figure 7.3 Load curves σ_m^* for a large bundle with thresholds following a Weibull distribution of index 2, with a strength reduction factor $a = 0.50$, and for different values of the maximum number m of damages before collapse.

7.2 Creep Failure

Creep is the deforming and subsequent weakening of a material caused by a long-term exposure to a constant applied load. The stress values that cause structural damage by creep failure may be considerably less than the yield strength of the material. Glacier flow is an example of creep processes in ice. Temperature is an important parameter, and creep is more pronounced at elevated temperatures.

Whereas fatigue is the weakening of the strength of a material being subjected to loading–unloading cycles up to a given stress level below the yield threshold, creep is time-dependent deformation due to a gradual increase in strain over time of a material subjected to constant stress below the yield threshold. We show in Figure 7.4 qualitatively how the deformation x evolves with time t when a load σ_0 has been applied. After a primary deformation phase, where the deformation rate goes down with time, there is a secondary phase where the deformation rate, $\mathrm{d}x/\mathrm{d}t$, is constant. The last, tertiary phase, is characterized by an increasing deformation rate. This phase ends with the material failing.

Typically, the secondary creep phase will last much longer than the other two and the lifetime t_f will be given by

$$\tau_\mathrm{f} \propto \frac{1}{(\mathrm{d}x/\mathrm{d}t)_2}. \tag{7.15}$$

where $(\mathrm{d}x/\mathrm{d}t)_2$ is the constant deformation rate during secondary creep. This deformation rate is described by the equation

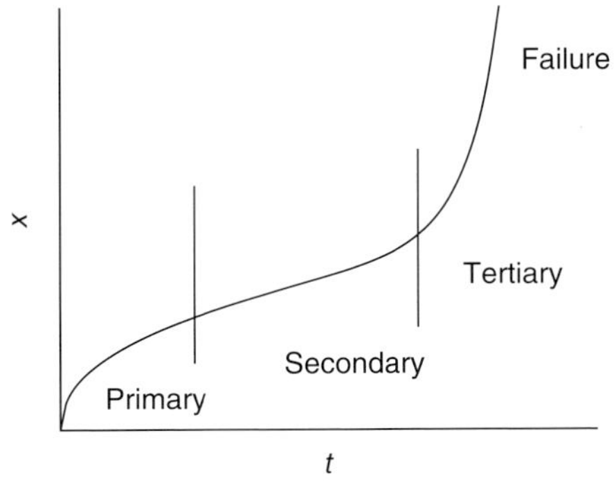

Figure 7.4 Deformation x as a function of time t due to creep.

$$\left(\frac{\mathrm{d}x}{\mathrm{d}t}\right)_2 = K\sigma_0^n e^{-Q/kT}, \tag{7.16}$$

where K is the creep constant, Q is the activation energy for creep, k the Boltzmann constant, T the temperature, and n the creep exponent. For metals, n is typically between 3 and 8 with 5 as the most common value.

Creep models are usually based on mechanisms leading to plasticity. The microcracks that appear are caused by creep. They affect the process, and in the tertiary creep phase, they may take over.

7.2.1
A Model for Creep

In this section, we present a simple model for creep. We model this scenario using a fiber bundle model. It may be an equal-load-sharing, a local-load-sharing or a soft clamp fiber bundle model. For this section, we base our discussion on the equal-load-sharing fiber bundle model in order to keep it simple.

We let the individual fibers in the fiber bundle model be responsible for the plasticity, whereas the failure of the fibers constitute the appearance of microcracks.

Let us start by modeling the response of a single fiber to a load σ. At any time t, fiber i will obey the Hooke law,

$$\sigma_i(t) = k_i(t)x(t). \tag{7.17}$$

The time dependency of the spring constant $k_i(t)$ represents the plasticity of the individual fibers.

We model the time evolution of $k_i(t)$ as

$$k_i(t) = \min_{t' \leq t}(1 - e^{-\eta_i(t')/\sigma_i(t')}) , \tag{7.18}$$

where $\eta_i(t)$ is a uniform noise on the unit interval, so that $\langle \eta_i(t) \rangle = 1/2$. $k_i(t)$ is to contain all the changes in the fiber due to plastic flow, the evolution of microcracks in the fiber, changes at the molecular level – in short, the aging of the material.

We expect this aging to have a stochastic component as temperature plays a role, see Eq. (7.16), hence $\eta_i(t)$. We must have that the aging of the material speeds up if the load $\sigma_i(t)$ is increased. We also note that for a given fixed load σ_i, $k_i(t) \geq 1 - \exp(-1/\sigma_i)$ for all t. This is so as $0 \leq \eta_i(t) \leq 1$. If $\sigma_i(t)$ decreases with time, the smallest value of $k_i(t')$ for $t' \leq t$ is retained afterward: there is no healing of the damage due to creep.[1)]

We now connect N fibers between two stiff clamps, thus obtaining an equal-load-sharing fiber bundle. The fibers have failure thresholds x_i drawn from some probability distribution $p(x)$. If the deformation – elongation of the fibers – x reaches the threshold value x_i of fiber i, this fiber fails. The total force on the fiber bundle is $F = \sigma N$. This force we keep constant. Hence, from Eq. (7.17), we have

$$F = x(t) \sum_{i=1}^{N} k_i(t), \tag{7.19}$$

so that the deformation is given by

$$x(t) = \frac{F}{\sum_{i=1}^{N} k_i(t)}. \tag{7.20}$$

The model is now easily implemented numerically. We discretize time by setting $t_n = n$. At $n = 0$, all $k_i(0) = 1$. We calculate $x(0)$ from Eq. (7.20). Then we check for each fiber i whether $x(0) \geq x_i$. If this is true, the corresponding fiber is removed. For the remaining fibers, we calculate the load $\sigma_i(0)$ from Eq. (7.17). We then generate N random numbers $\eta_i(0)$. From these random numbers together with the loads $\sigma_i(0)$, we calculate $k_i(1) = 1 - \exp[-\eta_i(0)/\sigma_i(0)]$. The spring constants $k_i(1)$ are then inserted into Eq. (7.19) to determine $x(1)$. We then check whether $x(1) \geq x_i$ for all remaining fibers, removing those where this is true and so on.

We show in Figure 7.5 the elongation $x(t)$ as a function of t for $F = 25$, 50, and 100. Comparing with Figure 7.4, we see that the secondary and tertiary creep phases are well reproduced. However, the hardening (i.e., the slowing down of $x(t)$) in the primary creep phase is not described by the model. Anything else would, however, not be possible. The spring constants $k_i(t)$ can only decrease with time, making hardening impossible within the model. We could, however, combine the present model with the modeling of work hardening that was presented in Section 7.1.

In Figure 7.6, we show a rescaling of the data in Figure 7.5. We find data collapse for both the secondary and tertiary creep phases with time rescaled by σ^2. We also find that $(dx/dt)_2$ to be constant during the secondary creep period. Hence, the creep Eq. (7.16) is obeyed with a creep exponent $n = 2$. This is lower than what is observed experimentally. However, by generalizing Eq. (7.18) to

$$k_i(t) = \min_{t' \leq t} (1 - e^{-\eta_i(t')/\sigma_i(t')})^{\gamma}, \tag{7.21}$$

1) A healing time t_h can be included by replacing the expression $\min_{t' \leq t}$ in Eq. (7.18) by $\min_{t-t_h \leq t' \leq t}$.

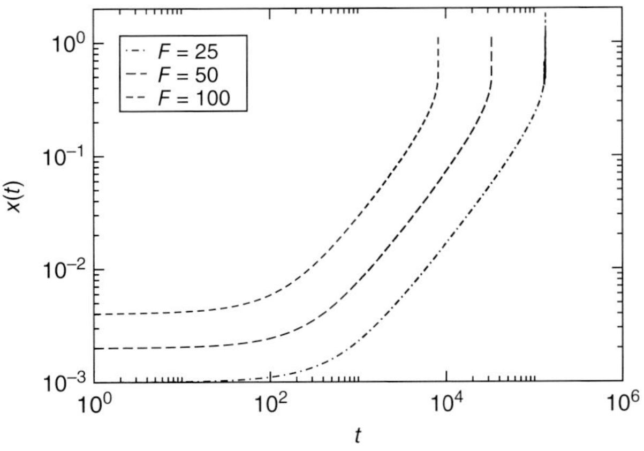

Figure 7.5 Elongation $x(t)$ as a function of t for different values of $\sigma = F/N$. The figure is based on 500 samples for $F = 25$ and one sample for $F = 50$ and 100. $N = 25\,000$.

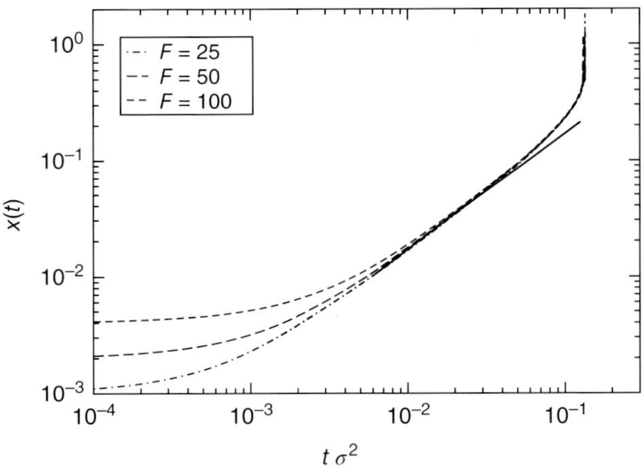

Figure 7.6 Elongation $x(t)$ as a function of $t\sigma^2$ for different values of $\sigma = F/N$. The straight line is $x = 1.7\sigma^2 t$. The figure is based on the data from Figure 7.5.

where γ is an exponent somewhat similar to the one introduced in the next section. We find the relation

$$n = 1 + \frac{1}{\gamma}. \tag{7.22}$$

We plot in Figure 7.7 lifetime τ_f versus σ, and find

$$\tau_f \sim \frac{1}{\sigma^2}, \tag{7.23}$$

consistent with Eq. (7.15).

The dynamics of the creep process proceeds through avalanches. We define the size of an avalanche as the number of fibers that fail between each consecutive quiescent silent period. Operationally, we do the following to measure the avalanche

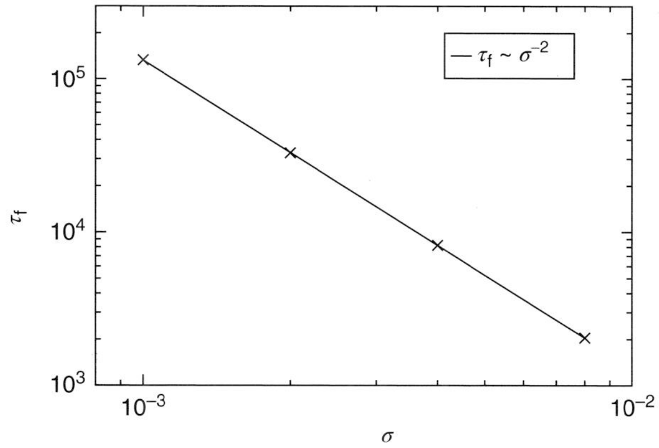

Figure 7.7 Lifetime τ_f versus σ for different threshold distributions using the equal-load-sharing model. Here $N = 25\ 000$ and 5 samples were generated for each σ. We used a threshold distribution that was flat on the unit interval.

distribution. We start summing the number of fibers that fail during each time step from the moment only one fiber fails during a time step. We stop summing when the number of fibers that fail during a time step again is one. The sum, Δ, is an avalanche.

In Figure 7.8, we show the histogram N_B of the avalanche sizes recorded during the creep process. We find that the power law

$$D(\Delta)\Delta^{-2.5}, \tag{7.24}$$

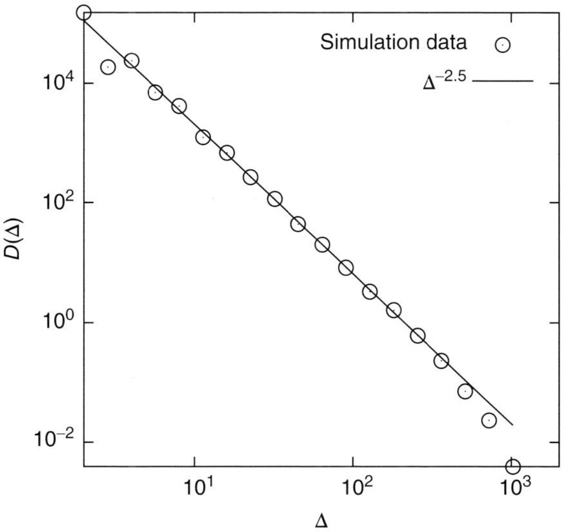

Figure 7.8 Histogram of the number of fibers that simultaneously fail – avalanches – throughout the failure process. The data are based on $N = 25\ 000$ and 500 samples. The threshold distribution was flat on the unit interval.

fits the date well over two and a half decades.

We have in this short section introduced a model for creep based on the fiber bundle model. Through its construction, the explicit modeling of microcracks in the system is done through failing fibers, whereas the complex rheology involved is modeled through the constitutive equations for the fibers. Even though the model is sketchy and not well developed as it is presented in this book, it may be turned into a useful tool by skilled readers.

7.2.2 A Second Model for Creep

We now turn to another fiber bundle model of creep, due to Kun *et al.* [88, 89]. In this model, a fiber may break for two different reasons. It may fail when the load exceeds the breaking threshold x_i of the fiber, and the fiber may also break when it has suffered sufficient damage $c(t)$. In this model, the damage $c(t)$ is a functional of all deformations $x(t')$ for $t' \leq t$.

7.2.2.1 Damage Accumulation

We consider first the standard equal-load-sharing fiber bundle subjected to a constant load. The amount of damage, $c(t)$, on a fiber depends on the deformation $x(t)$ it has experienced. One may assume that the damage $\mathrm{d}c$ inflicted in a small time interval $\mathrm{d}t$ is of the form

$$\mathrm{d}c(t) = ax(t)^{\gamma}\ \mathrm{d}t, \tag{7.25}$$

parameterized by the two positive constants a and γ. Hence, the total accumulated damage at time t is an integral over the loading history:

$$c(t) = a\int_0^t x(t')^{\gamma}\ \mathrm{d}t'. \tag{7.26}$$

A fiber can only tolerate a finite amount x_i of load and a finite amount c_i of damage. If either of these thresholds is exceeded, the fiber breaks. The two breaking strengths may be assumed to be statistically uncorrelated with cumulative distribution functions $P_{\mathrm{d}}(c)$ and $P(x)$ for the breaking thresholds c_i and x_i, respectively. In spite of the thresholds being uncorrelated, the two breaking modes are dynamically coupled because damage accumulation beyond the threshold c_i causes the fiber number i to fail, which in turn increases the load on the intact fibers.

The force per fiber, σ, is given by

$$\sigma = [1 - P_{\mathrm{d}}(c(t))][1 - P(x(t))]x(t), \tag{7.27}$$

generalizing the load curve (2.13), $\sigma = [1 - P(x)]x$, for equal load sharing in the absence of fatigue. The stress σ is kept constant here. A natural choice for the statistical distribution of the damage accumulation thresholds is a uniform density, for simplicity taken on the interval $(0, 1)$, corresponding to the cumulative distribution $P_{\mathrm{d}}(c) = c$. Then Eq. (7.27) takes the form

$$c(t) = 1 - \frac{\sigma}{[1 - P(x)]x}, \tag{7.28}$$

where $x = x(t)$.

7.2.2.2 Time Evolution

One may study the time evolution $x(t)$ in the beginning of the degradation process when immediate failures hardly occur, that is, $P(x(t)) \approx 0$. The time evolution in this limiting case is then given by the solution of

$$\sigma = [1 - c(t)]x(t), \tag{7.29}$$

together with damage accumulation law (7.25) or (7.26). Differentiating $1 - c(t) = \sigma/x(t)$ with respect to time we have

$$\frac{\sigma}{x^2} \cdot \frac{dx}{dt} = \frac{dc}{dt} = a\, x^{\gamma}, \tag{7.30}$$

by means of (7.25). Thus

$$x^{-\gamma-2}\frac{dx}{dt} = a/\sigma \quad \text{or} \quad \frac{d}{dt}x^{-\gamma-1} = -(1+\gamma)a\sigma^{-1}. \tag{7.31}$$

Integration yields

$$x^{-\gamma-1} = a(1+\gamma)\sigma^{-1}(t_f - t), \tag{7.32}$$

where t_f is an integration constant. If the process starts before any damage accumulation occurs, that is, $x(0) = \sigma$, the integration constant must satisfy

$$t_f = \frac{1}{a(1+\gamma)\sigma^{\gamma}}. \tag{7.33}$$

Consequently, the time evolution is as follows:

$$x(t) = \sigma[1 - t/t_f]^{-1/(1+\gamma)}. \tag{7.34}$$

Thus, the damage accumulation leads to a finite time singularity where the extension $x(t)$ of the bundle has a power-law divergence. The lifetime of the system is, therefore, t_f. This is the so-called exact solution of Kun *et al.* [89].

When one is *not* near the start of the degradation process, the effect of the broken fibers must be taken into account, and the complete Eq. (7.28) must be considered. Differentiating (7.28) with respect to time, we find

$$\frac{dc}{dt} = \sigma\,\frac{dx}{dt}\,Q(x), \quad \text{where } Q(x) = -\frac{d}{dx}\left[\frac{1}{[1-P(x)]x}\right]. \tag{7.35}$$

As $dc/dt = ax(t)^{\gamma}$ is the damage accumulation per unit time, one can eliminate $c(t)$ to obtain a relation between x and time:

$$dt = \frac{\sigma}{a}\, x^{-\gamma}\, Q(x)\, dx. \tag{7.36}$$

Integration yields

$$t_f - t = \frac{\sigma}{a}\int_x^{x_c} y^{-\gamma}\, Q(y)\, dy, \tag{7.37}$$

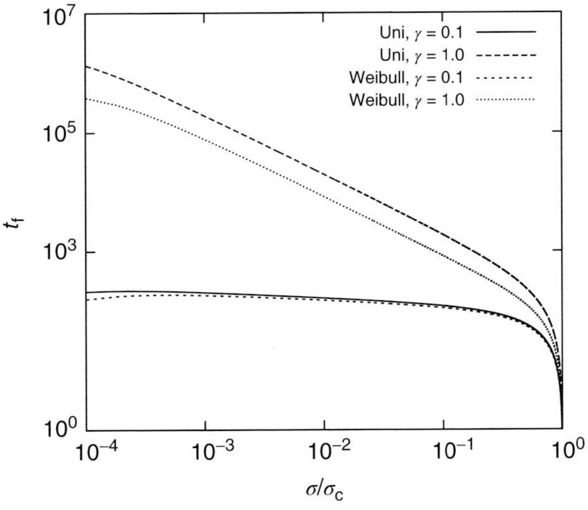

Figure 7.9 Lifetimes as function of the external load σ for fiber bundles with uniform and with Weibull, index 5, strength distributions. The slope of the graphs at low load equals the damage accumulation parameter γ.

where t_f is an integration constant. The critical point x_c corresponds to t_f.

One sees that t=0 corresponds to

$$t_f = \frac{\sigma}{a} \int_{x(0)}^{x_c} y^{-\gamma} Q(y) \, dy. \tag{7.38}$$

Assuming that one starts with no damage, $x(0)$ and σ are connected through the load curve for the undamaged bundle, $\sigma = x(0)[1 - P(x(0))]$. This gives the desired connection between the lifetime and the applied stress σ.

Figure 7.9 shows numerical result for the lifetime t_f for two bundles with breaking thresholds x_i distributed according to a uniform distribution and to a Weibull distribution of index 5.

The time-dependent deformation $x(t)$ is determined by Eq. (7.37). Figure 7.10 shows that the deformation changes slowly for a considerable time, but accelerates quickly in the last stage, just before complete breakdown.

7.2.2.3 **Healing**

Experiments have shown that healing of microcracks in the material plays an important part in creep processes, especially at low loads. Healing can be captured in the fiber bundle model by introducing a finite-ranged memory time τ during which the loading history contributes to the accumulated damage. Assuming exponentially decaying memory, one may replace the expression (7.26) for the damage accumulation by

$$c(t) = a \int_0^t e^{-(t-t')/\tau} x(t')^\gamma \, dt'. \tag{7.39}$$

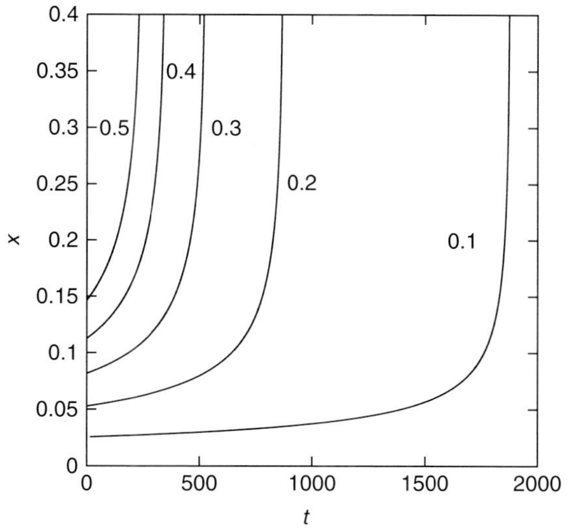

Figure 7.10 Extension of fibers as function of time, for different values of the load σ/σ_c. The fiber thresholds are uniformly distributed on the unit interval.

As now

$$\frac{\mathrm{d}c}{\mathrm{d}t} = ax(t)^{\gamma} - \frac{c(t)}{\tau}, \tag{7.40}$$

the healing clearly reduces the damage accumulation.

Healing becomes dominating when the memory range τ becomes comparable with the lifetime t_f of the system without healing.

7.3 Viscoelastic Creep

Let us assume that a bundle consists of fibers that are not merely elastic, but *viscoelastic*. Then the stress depends both on the size of the strain x and on how rapidly the strain changes. For a single fiber, the stress–strain relation takes the form

$$\sigma = \kappa x + \beta \dot{x}, \tag{7.41}$$

where β is a damping coefficient, and κ is the elastic constant. As before, we set κ equal to unity for all fibers.

Assuming equal load sharing, the stress–strain relation for a large bundle with a cumulative threshold distribution $P(x)$ takes the form

$$\sigma = (\beta \dot{x} + x)[1 - P(x)], \tag{7.42}$$

a straightforward generalization of the load curve (2.13) for a bundle of elastic fibers. The complete description of the time-dependent process is contained in this differential equation.

If the strain $x(t)$ converges toward a value x_s, this stationary value satisfies Eq. (7.42) with $\dot{x} = 0$:

$$\sigma = x_s[1 - P(x_s)]. \tag{7.43}$$

As the load curve $x[1 - P(x)]$ for the elastic fibers has a maximum σ_c at a critical point x_c, the stationarity condition (7.43) corresponds to two fixed points for the system if $\sigma < \sigma_c$ and no real fixed point for $\sigma > \sigma_c$.

To survey the nature of the differential Eq. (7.42), we plot $\dot{x}$ as a function of the strain x,

$$\beta\dot{x} = \frac{\sigma}{1 - P(x)} - x. \tag{7.44}$$

The general behavior will be as in Figure 7.11. For $\sigma < \sigma_c$, there will be the two fixed points x_s satisfying (7.43); for $\sigma = \sigma_c$, one fixed point where the graph touches the horizontal axis; and for $\sigma > \sigma_c$, no fixed points.

When the stress is close to the critical value, the velocity $\dot{x}$ will be very small in a region near the critical point $x = x_c$. Thus, the process will proceed very slowly near criticality, for both precritical and postcritical stress σ.

For $\sigma < \sigma_c$ we see from Figure 7.11 that the smaller fixed point is attractive, while the other one is repulsive. Thus, the system ends up at the smaller fixed point as long as the initial condition $x(0)$ is less than the repulsive fixed point.

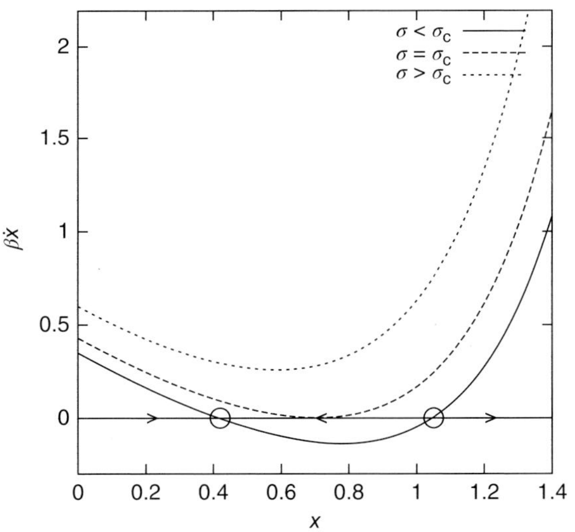

Figure 7.11 The general behavior of the strain change versus the strain x, according to Eq. (7.42). The dashed graph corresponds to the critical stress σ_c, and the intersection of a graph with the ordinate axis gives the stress value. The arrows indicate the flow direction along the x axis. The fixed points are marked as circles. The graphs in the figure correspond to a Weibull threshold distribution of index 2, $P(x) = 1 - \exp(-x^2)$.

In order to see more quantitatively how the stationary state x_s is approached for $\sigma < \sigma_c$, we introduce the deviation,

$$\delta(t) = x_s - x(t), \tag{7.45}$$

and expand the differential equation (7.42) to linear order in $\delta(t)$. Using (7.43) and $P(x_s + \delta) = P(x_s) + p(x_s)\,\delta + \mathcal{O}(\delta^2)$, we obtain to dominating order

$$\beta[1 - P(x_s)]\frac{\mathrm{d}\delta}{\mathrm{d}t} = x_s[1 - P(x_s) - x_s p(x_s)]\delta. \tag{7.46}$$

Thus, $\delta(t)$ decays exponentially,

$$\delta(t) \propto e^{-t/\tau}, \tag{7.47}$$

with

$$\tau = \beta\frac{1 - P(x_s)}{1 - P(x_s) - x_s p(x_s)}. \tag{7.48}$$

τ is the characteristic time of the relaxation process.

We know from (2.14) that the denominator vanishes at the critical point $x_s = x_c$, so that

$$\tau \propto (x_c - x_s)^{-1} \tag{7.49}$$

Hence, the relaxation time diverges at the critical point. The load curve has a parabolic maximum at criticality,

$$\sigma = \sigma_c - a(x_c - x)^2, \tag{7.50}$$

where a is given by the threshold distribution, (see Eq. (3.15)). Consequently, we have

$$\tau \propto (\sigma_c - \sigma)^{-1/2} \tag{7.51}$$

in terms of the stress. Thus, the relaxation time diverges near the critical stress with an exponent $-1/2$.

How does the system behave for postcritical stress? In this case, there exists no stationary situation, so that the strain increases at all times, as Figure 7.11 shows. Large stress accelerates the process, while it proceeds slower near the critical stress. Let us obtain an estimate of the near-critical delay.

It is clear from the figure that the velocity behaves as

$$\frac{\mathrm{d}x}{\mathrm{d}t} \propto (x - x_c)^2 \tag{7.52}$$

for σ near σ_c. Hence, the integration picks up a contribution

$$t_d \propto \frac{1}{|x - x_c|} \tag{7.53}$$

in a region near x_c. As the stress depends quadratically on $x - x_c$ near criticality (see the figure),

$$\sigma - \sigma_c \propto (x - x_c)^2, \quad \text{i.e. } |x - x_c| \propto (\sigma - \sigma_c)^{1/2}, \tag{7.54}$$

the delay (7.53) takes the form

$$t_d \propto (\sigma - \sigma_c)^{-1/2} \qquad (7.55)$$

near the critical point.

Thus, the characteristic time for the creep process has a two-sided power-law divergence at σ_c with a universal exponent $-1/2$ independent of the specific form of the threshold distribution $P(x)$. Note the similarity to the universal relaxation times (5.120) for elastic fiber bundles.

7.4
Fatigue Failure

Fatigue is the weakening of a material being subjected to repeated loading and unloading. A simple example is breaking of a thin steel rod with your hands after bending it back and forth several times in the same place. The stress values that cause structural damage – fatigue failure – may be considerably less than the normal yield of the material.

In practical cases, as well as in laboratory testing, the stresses are caused by cyclic loading. For cyclic loading, the deformation is expected to diverge after a number N_f of cycles. Empirically, the necessary number of cycles, the life time t_f, decreases with the loading amplitude σ_0 as a power law,

$$t_f \propto \sigma_0^{-\beta}. \qquad (7.56)$$

This power-law dependence (7.56) is known as Basquin's law. It is found experimentally in a broad class of materials [85]. The values of the Basquin exponent β depend on the material properties and vary over a broad range. For ductile materials like metals and alloys, the exponent is of the order 0.1, for the asphalt experiment described in Section 7.4 $\beta \approx 2.2$, and a protein gel experiment provides a value as high as $\beta \simeq 5.5$ [90].

Let us consider an interesting example.

7.4.1
A Fatigue Experiment

Kun *et al.* [88] exposed a cylindrical asphalt specimen to cyclic diametrical compressions at an external load of constant amplitude σ_0. To connect the experiment with a fiber bundle setting, one may consider the cylinder as two half cylinders held together by asphalt fibers (Figure 7.12). The experiment proceeds until a macroscopic crack appears along the load direction.

Figure 7.13 shows the experimental deformation $x(t)$ as a function of the number N_{cycle} of loading cycles, for two different values of the loading amplitude σ_0. In the figure, it is compared with theoretical results by Kun *et al.*, who used the framework developed for creep in Section 7.2.2. These are obtained by numerical integration of (7.39), where now the stress σ is no longer constant, but

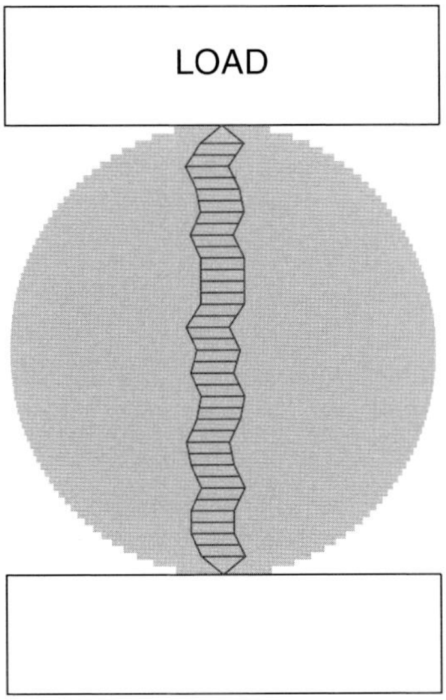

Figure 7.12 The asphalt sample modeled as a fiber bundle.

time-dependent, representing the cyclic loading. The damage accumulation $c(t)$ includes healing. The agreement is excellent.

Figure 7.14 shows the dependence of the number N_f of cycles that give total collapse in the experiment as a function of the load amplitude σ_0. It is compared with the numerical results for a fiber bundle with a Weibull fiber strength distribution $P(x)$. The fit is excellent, except for very large amplitudes. In the large middle range of loading amplitudes, Basquin's exponential law (7.56) is well obeyed.

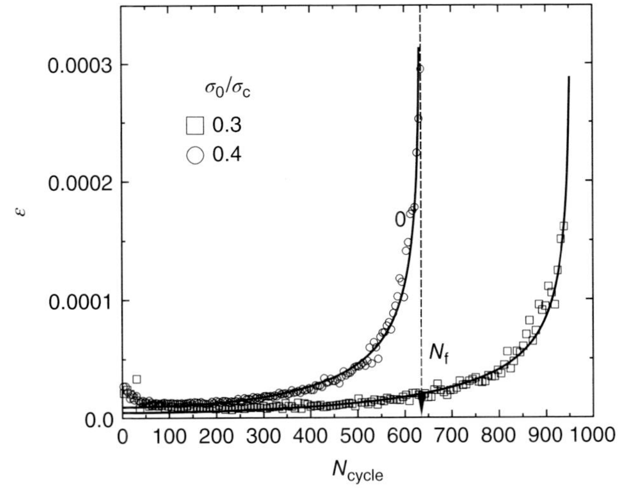

Figure 7.13 Experimental deformation $x(t)$ as a function of the number of cycles N_{cycle}. The continuous line is a fit with theoretical fiber bundle results, assuming a uniform distribution of thresholds x_i. From Ref. [88].

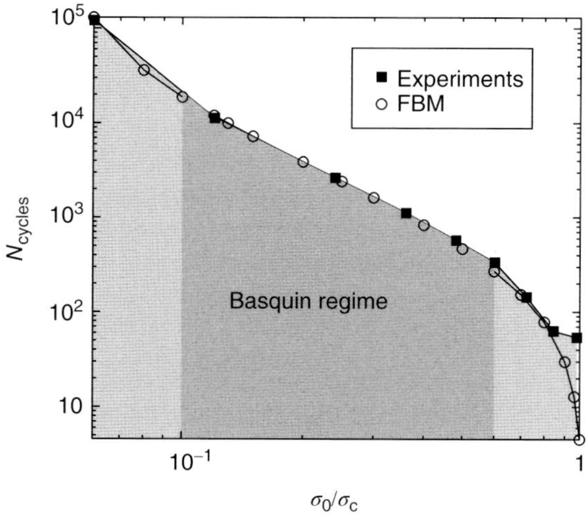

Figure 7.14 The number N_f of cycles causing total failure as a function of the load amplitude σ_0. Both experimental and numerical fiber bundle results are shown. The fibers have Weibull distributed strengths, damage accumulation parameters $\gamma = 2, a = 0.01$, and healing range $\tau = 15\,000$. From Ref. [88].

7.5 Thermally Induced Failure

Thermal fluctuations induce a Gaussian noise in loaded materials and may, therefore, cause failure. This will now be demonstrated in a fiber bundle with equal load sharing [91].

7.5.1 Failure Time for a Homogeneous Fiber Bundle

We assume for simplicity that the bundle is homogeneous so that all fibers have the same breaking strength σ_c. The stress on each fiber is

$$\sigma = \sigma_0 + \eta, \tag{7.57}$$

where σ_0 is the applied external stress F/N, and η is a random noise with a Gaussian distribution of zero mean and variance kT:

$$p(\eta) = \frac{1}{\sqrt{2\pi kT}} \exp\left(-\frac{\eta^2}{2kT}\right). \tag{7.58}$$

The probability that the noise value exceeds η equals

$$P(\eta) = \int_{\eta}^{\infty} p(y)\ \mathrm{d}y, \tag{7.59}$$

so that $1 - P$ is the usual cumulative distribution function.

The probability that a single fiber survives at an arbitrary instant of time equals $1 - P(\sigma_c - \sigma_0)$. Hence, the probability that a fiber survives at time t can be written

$$p_1(t) = [1 - P(\sigma_c - \sigma_0)]^t, \tag{7.60}$$

which defines a time scale.

Then the probability that the entire bundle survives at time t is

$$p_N(t) = [1 - P(\sigma_c - \sigma_0)]^{Nt} = e^{-t/\langle\tau_1\rangle}, \tag{7.61}$$

with the average first failure time

$$\langle\tau_1\rangle = \frac{-1}{N \ln [1 - P(\sigma_c - \sigma_0)]}. \tag{7.62}$$

After the first fiber has failed, the situation remains the same with a smaller bundle and larger stress. Thus, the average failure time after $i - 1$ broken fibers is

$$\langle\tau_i\rangle = \frac{-1}{(N - i) \ln \{1 - P[\sigma_c - N\sigma_0/(N - i)]\}}. \tag{7.63}$$

Thus, the average total failure time $\langle\tau_f\rangle$ for the bundle is obtained by summing all these failure times:

$$\langle\tau_f\rangle = \sum_{i=0}^{N-1} \frac{-1}{(N - i) \ln \{1 - P[\sigma_c - N\sigma_0/(N - i)]\}}. \tag{7.64}$$

When N is large, one can replace this sum by an integral,

$$\begin{aligned}\langle\tau_f\rangle &= \int_0^N \frac{-\,\mathrm{d}y}{(N - y) \ln \{1 - P[\sigma_c - N\sigma_0/(N - y)]\}} \\ &= \int_{\sigma_0}^{\infty} \frac{-1}{\ln [1 - P(\sigma_c - z)]} \frac{\mathrm{d}z}{z},\end{aligned} \tag{7.65}$$

by introducing $y = N - N\sigma_0/z$. Equation (7.65) is exact for a large homogeneous fiber bundle.

7.5.2
Failure Time for Low Thermal Noise

When the load σ_0 is smaller than, the maximum load σ_c that a fiber can sustain, and the temperature is sufficiently low, $kT < \sigma_c - \sigma_0$, one can achieve a closed-form expression for the average failure time. By taking into account that the above sum is dominated by the time required for breaking the first fiber, the probability P will be much smaller than 1. The derivative of $\langle\tau_f\rangle$ with respect to σ_0 gives

$$\frac{\mathrm{d}\langle\tau_f\rangle}{\mathrm{d}\sigma_0} = \frac{1}{\sigma_0 \ln [1 - P(\sigma_c - \sigma_0)]} \simeq \frac{1}{\sigma_0 P(\sigma_c - \sigma_0)} \tag{7.66}$$

Thus, we need the cumulative distribution

$$P(\sigma_c - \sigma_0) = \int_{\sigma_c - \sigma_0}^{\infty} \sqrt{\frac{1}{2\pi kT}} \exp(-\eta^2/2kT)\, \mathrm{d}\eta = \frac{1}{\sqrt{2\pi}} \int_{\xi}^{\infty} e^{-y^2/2}\, \mathrm{d}y, \tag{7.67}$$

with

$$\xi = \frac{\sigma_c - \sigma_0}{\sqrt{kT}}, \tag{7.68}$$

for large arguments ξ. Introducing the new integration variable z through

$$y = \xi + z/\xi, \tag{7.69}$$

we have for large ξ

$$\int_{\xi}^{\infty} e^{-y^2/2}\, \mathrm{d}y = \frac{e^{-\xi^2}}{\xi} \int_0^{\infty} e^{-z}\, e^{-z^2/2\xi^2}\, \mathrm{d}z \simeq \frac{e^{-\xi^2/2}}{\xi}, \tag{7.70}$$

as the factor $\exp\,(-z^2/2\xi^2)$ is asymptotically equal to 1.

Using (7.67), (7.68), and (7.70), we then have

$$\begin{aligned} \frac{\partial \langle \tau_f \rangle}{\partial \sigma_0} &= \sqrt{2\pi/kT}(\sigma_c - \sigma_0)\sigma_0^{-1} e^{(\sigma_c-\sigma_0)^2/2kT} \\ &= \frac{\partial}{\partial \sigma_0}\left[\sqrt{2\pi kT}\sigma_0^{-1} e^{(\sigma_c-\sigma_0)^2/2kT}\right] + \sqrt{2\pi kT}\sigma_0^{-2} e^{(\sigma_c-\sigma_0)^2/2kT}. \end{aligned} \tag{7.71}$$

The last term is small, of relative order kT, so for the dominating order, we end up with

$$\langle \tau_f \rangle = \frac{\sqrt{2\pi kT}}{\sigma_0} \exp\left(\frac{(\sigma_c - \sigma_0)^2}{2kT}\right). \tag{7.72}$$

The exponential expression (7.72) shows that the average failure time is very long under the assumed conditions. More precisely, we have the following dominating behavior:

- For fixed external load: $\ln\,\langle \tau_f \rangle \simeq 1/(kT)$.
- For fixed temperature: $\langle \tau_f \rangle$ diverges as $1/\sigma_0$ when the external load σ_0 approaches zero.

For an inhomogeneous fiber bundle in which the fibers have a distribution of strengths, the analysis is more difficult. We merely note that numerical simulations [92] indicate that the average failure time is essentially given by the same expression, provided that the temperature T is replaced by a higher effective temperature T_{eff}, as if the heterogeneities in the medium provide an additional noise.

7.6 Noise-Induced Failure

Let us assume that the noise level, whatever sources of noise there are present, can be characterized by a single dimensionless noise parameter $\hat{T}$. In general, we expect the noise to be a small perturbation of the system, $\hat{T}$ small, but its long-time effects may nevertheless be important.

Under the influence of noise in addition to the external stress σ, a fiber will have a probability

$$p_f(\sigma_{\text{eff}}, \hat{T}) \tag{7.73}$$

of failing. The effective stress σ_{eff} on the fiber depends on the fraction n_t, the relative number of surviving fibers at the time,

$$\sigma_{\text{eff}} = \frac{\sigma}{n_t}. \tag{7.74}$$

Without precise information about the noise mechanisms, one cannot give the precise form of the failure probability. Let us, therefore, be satisfied with a simple qualitative expression for p_f, which consequently can be used to obtain a qualitative description of the bundle breakdown.

Which conditions should one expect p_f to satisfy? We must have the following restrictions:

- In the absence of noise, we must have

$$p_f(\sigma_{\text{eff}}, 0) = \begin{cases} 0 & \sigma_{\text{eff}} < x_i \\ 1 & \sigma_{\text{eff}} \geq x_i, \end{cases} \tag{7.75}$$

  where x_i is the breaking threshold of fiber i.
- p_f should increase with increasing stress σ_{eff}
- p_f should increase with increasing noise $\hat{T}$.

A simple expression satisfying these three conditions is

$$p_f(\sigma_{\text{eff}}, \hat{T}) = \begin{cases} A \exp\left[-\frac{1}{\hat{T}}\left(\frac{x_i}{\sigma_{\text{eff}}} - 1\right)\right] & \sigma_{\text{eff}} < x_i \\ 1 & \sigma_{\text{eff}} \geq x_i, \end{cases} \tag{7.76}$$

where A is a constant [93].

Let us now for simplicity consider a *homogeneous* bundle, in which all fibers have the same threshold, which therefore will be the critical threshold: $x_i = \sigma_c$ for all i, that is,

$$p_f(\sigma_{\text{eff}}, \hat{T}) = \begin{cases} A \exp\left[-\frac{1}{\hat{T}}\left(\frac{\sigma_c}{\sigma_{\text{eff}}} - 1\right)\right] & \sigma_{\text{eff}} < \sigma_c \\ 1 & \sigma_{\text{eff}} \geq \sigma_c, \end{cases} \tag{7.77}$$

The existence of a nonzero failure probability implies that the bundle sooner or later will break down. In order to determine the bundle lifetime, we note that the reduction rate of the number of intact fibers is given as:

$$\frac{\mathrm{d}n_t}{\mathrm{d}t} = -n_t\, p_f(\sigma_{\text{eff}}, \hat{T}). \tag{7.78}$$

The *failure time* τ_f, the time to reach complete breakdown, is the time it takes to reduce n_t from 1 to 0. Thus

$$\tau_f = \int_0^1 \frac{1}{n_t\, p_f(\sigma/n_t, \hat{T})}\, \mathrm{d}n_t. \tag{7.79}$$

The integral is not straightforward, but if one for analytic convenience takes $A = \sigma_{\rm eff}/\sigma_{\rm c}$, the integration is trivial [93]. One obtains

$$\tau_{\rm f} = \hat{T} \exp\left(-\frac{1}{\hat{T}}\right)\left[\exp\left(\frac{\sigma_{\rm c}}{\sigma_{\rm eff}\hat{T}}\right) - 1\right] \tag{7.80}$$

One sees that, in general, the failure time is exponentially long. However, when $\sigma_{\rm eff}$ approaches $\sigma_{\rm c}$, $\tau_{\rm f}$ approaches zero, which is reasonable.

We will now study the failure scenario in more detail, starting from an intact bundle. If the noise level is very low, or if the bundle stress initially is low, not much happens to the bundle. There will be a waiting time until the noise has done its effect, resulting in a fiber failure. Then either there is a new waiting time or the redistribution of the load carried by the failed fiber produces another fiber failure. The latter possibility leads to a situation with continuous ruptures. One may, therefore, say that the bundle can be in one of two phases, one phase with continuous fracturing (no waiting times) and another intermittent slow phase.

Let us define σ_0 as the external stress value for triggering continuous fracturing at a given noise level $\hat{T}$. Thus, the boundary between these phases is a curve in the $(\sigma_0, \hat{T})$ plane. We estimate that the boundary corresponds to the expectation that precisely one of the N fibers fails, that is,

$$N\, p(\sigma_0, \hat{T}) = 1, \tag{7.81}$$

or

$$N\, A \exp\left[-\frac{1}{T}\left(\frac{\sigma_{\rm c}}{\sigma_0} - 1\right)\right] = 1. \tag{7.82}$$

Consequently,

$$\sigma_0 \simeq \frac{\sigma_{\rm c}}{1 + \hat{T} \ln\,(N)}. \tag{7.83}$$

This dominating behavior is independent of the constant A.

In the absence of noise, this estimate gives $\sigma_0 = \sigma_{\rm c}$, as one could expect, as in this case, continuous fracturing only takes place under postcritical conditions. In Figure 7.15, the analytical boundary estimate (7.83) in the $(\hat{T}, \sigma_0)$ plane is shown and compared with Monte Carlo simulation results for a bundle with $N = 2000$ fibers.

Clearly, the waiting times $t_{\rm w}$ in the intermittent phase will take different values, with shorter waiting times near the phase boundary. The distribution $D(t_{\rm w})$ of waiting times, shown in Figure 7.16, can be numerically fitted to a gamma distribution,

$$D(t_{\rm w}) \propto \exp\,(-t_{\rm w}/a)/t_{\rm w}^{1-\gamma}, \tag{7.84}$$

with $\gamma \simeq 0.15$ and $a \simeq 800$ for homogeneous bundle.

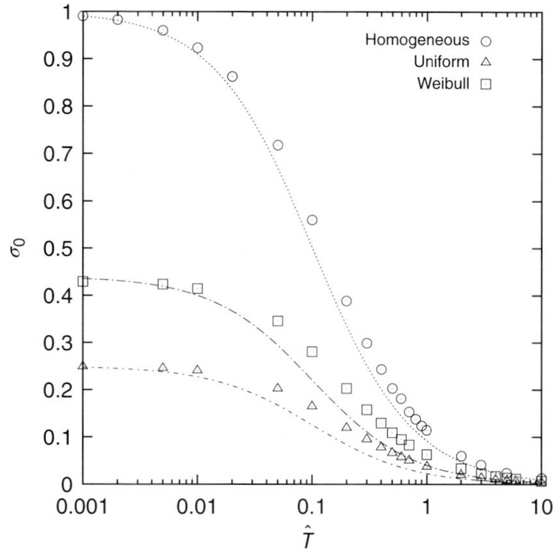

Figure 7.15 The phase boundary in the $(\hat{T}, \sigma_0)$ plane for three different types of fiber strength distributions. The data points show simulation results for a homogeneous bundle with $\sigma_c = 1$ (circles), for a uniform threshold distribution (triangles), and for a Weibull distribution of index 2 (squares). The lines are curves of the form (7.83). The number of fibers in each bundle was $N = 20\ 000$.

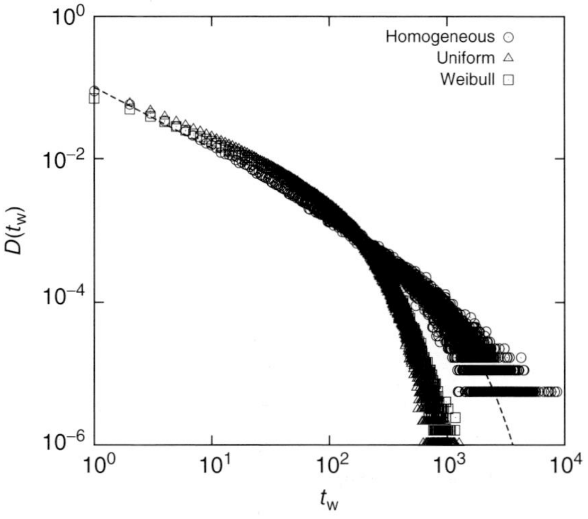

Figure 7.16 Simulation results for the waiting time distribution $D(t_w)$ for three different types of fiber strength distributions. The graphs can be fitted with gamma functions of the type (7.84), where $\gamma \simeq 0.15$ for a homogeneous bundle, and $\gamma \simeq 0.26$ for uniform and Weibull distributions. Each bundle contains $N = 20\ 000$ fibers.

So far, we have only considered a homogeneous bundle. For an inhomogeneous bundle, that is, with a statistical distribution of fiber threshold values, similar behavior is seen. Figure 7.16 shows simulation results for the phase boundary, for a uniform flat strength distribution, and for a Weibull distribution of index 2. The form is similar to the one for the homogeneous bundle.

For an inhomogeneous bundle, the slow intermittent phase may be preceded by a continuous fracturing phase until the quasistationary state, where $\sigma \simeq x[1 - P(x)]$, is reached. The distribution of the subsequent waiting times for the uniform and the Weibull threshold distributions may again be represented by the gamma function (7.84), with different parameters (see Figure 7.16).

Existence of a phase boundary of the type described has important consequences on breakdown phenomena in materials. In a real situation with material fracturing, acoustic emission measurements can show clearly whether an ongoing breakdown process belongs to a continuous fracturing or intermittent fracturing phase. Once a system enters a continuous fracturing phase, the total system failure will be imminent. Thus, the phase identification can give important warnings.

7.7
Crushing: The Pillar Model

Many failures occur as a result of compression. The *pillar* version of the fiber bundle model is a useful tool for modeling a number of such situations, see Figure 7.17.

In connection with nano- and micro-electromechanical systems, arrays of load-bearing nanopillars are fabricated [94, 95]. These pillars may be analyzed

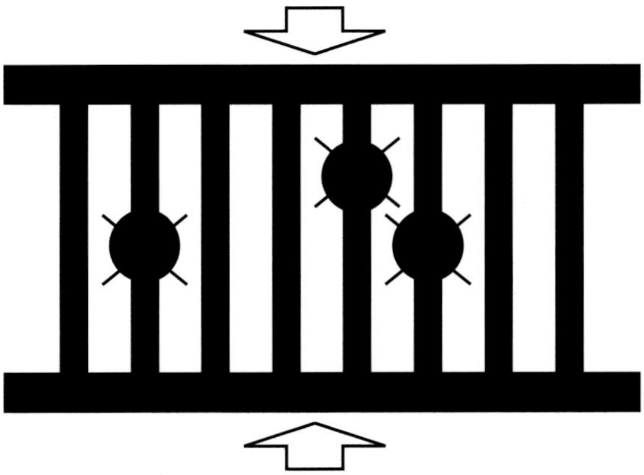

Figure 7.17 The pillar model. The fibers have been substituted by pillars that fail under stress. The pillars are loaded under compression, and they fail according to their failure thresholds.

as the fiber bundle model, the only difference being that the fibers now are pillars that are compressed rather than elongated. Domański *et al.* [15] used the two-dimensional local-load-sharing model (see Section 4.2) when studying this system.

Porous rocks consist of cemented grains. Under compressive loading, either the cementation between the grains fails (intergranular failure) or the grains themselves fail (intragranular failure). In Figure 7.18, we show a porous rock sample that has failed after being loaded along the long axis. The macroscopic failure has been localized to a plane.

The equal-load-sharing *pillar model* has been used to explain qualitative features of the failure mode seen in Figure 7.18 [16]. The plane in which the failure eventually occurs is modeled as an array of N pillars. The pillars fail one by one as the loading increases, at a rate of r pillars per strain unit. What will be the stress–strain relationship for this system – which then is assumed to be the stress–strain relation of the rock itself under these loading conditions? If k is the number of pillars that have been crushed, then the stress per pillar is

$$\sigma = \frac{E_0}{N}(N - k)x\,, \tag{7.85}$$

where E_0 is the stiffness of the pillars and x is the strain. We furthermore have that $k = rx$, giving

$$\sigma = \left(1 - \frac{rx}{N}\right) E_0 x. \tag{7.86}$$

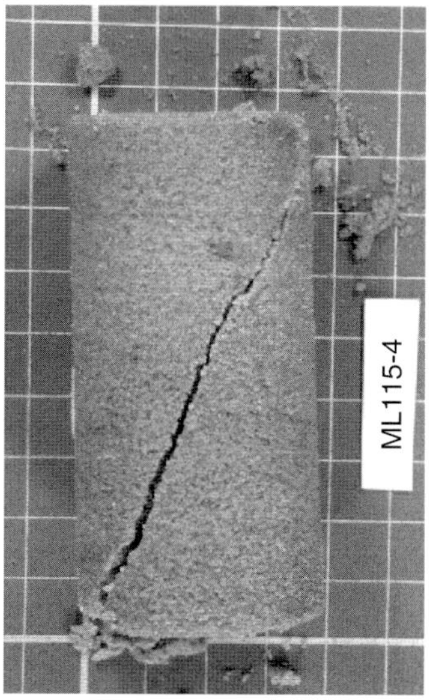

Figure 7.18 Photo of sandstone after being loaded uniaxially along the long axis. (Photo credit: SINTEF.)

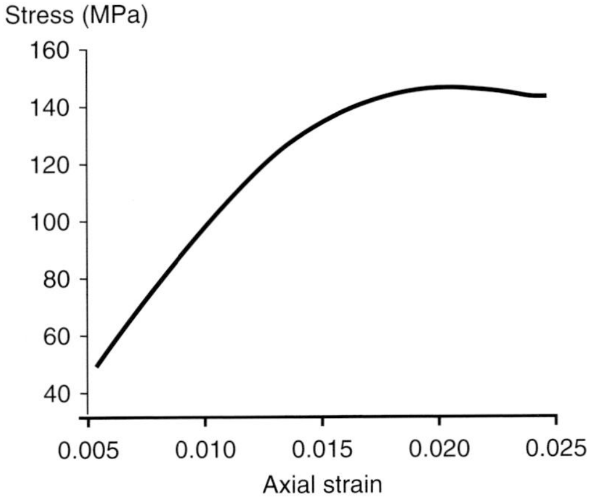

Figure 7.19 Stress–strain relation for Red Wildmoor sandstone as determined in a triaxial test. (Curve taken from Ref. [96].)

This is the stress–strain relation in the pillar model. We show in Figure 7.19 the stress–strain curve of Red Wildmoor sandstone as measured in a triaxial test [96]. This curve is well fitted by a parabola up to the stress maximum.

8
Snow Avalanches and Landslides

The subtitle of this book associates fiber bundles with the task of modeling failure in *materials*. As shown in the current literature, "materials" must be taken in a very broad sense.

In this chapter, we describe briefly two unconventional applications, namely how fiber bundle models have been used to describe aspects of natural phenomena in which rapid mass movements are triggered. We have seen in Chapter 3 how avalanches in the model could be triggered by small causes, and it is, therefore, not unreasonable to envisage that fiber bundle models may be applied to important natural phenomena of similar kind: avalanches of snow and of earth. Such applications can be found in the current literature. In Section 8.1, we report on a snow avalanche modeling, and in Section 8.2, we explain briefly how modeling of landslides has been set up.

8.1
Snow Avalanches

The most dangerous snow avalanches are *slab avalanches* in which, at least initially, a complete sheet of snow starts sliding. Further down the slope, the slab may break up into smaller pieces. Loose-snow avalanches do also occur, but 99% of fatal avalanches in Canada during the period 1972–1991 were slab avalanches.

A slab snow avalanche presupposes the existence of a *weak layer* below the surface, usually caused by a hoarfrost[1] layer created before later snowfalls with dry snow. A slab above the weak layer may be prevented from sliding by snow grains that connect across the layer. These bonding snow grains are modeled as fibers extending across the weak layer (Figure 8.1) [97, 98].

The load on the inclined slab, caused by gravity, will macroscopically be seen as a shear strain (Figure 8.2). The upper plate moves orthogonally to the initial fiber direction. Through the shear strain, the fibers are stretched and may rupture. The thresholds for the rupture strengths for the individual fibers are assumed to be given by a statistical Weibull distribution.

1) Hoarfrost is a crystalline frost that forms from deposition of moisture from low-humidity air.

The Fiber Bundle Model: Modeling Failure in Materials, First Edition.
Alex Hansen, Per C. Hemmer and Srutarshi Pradhan.

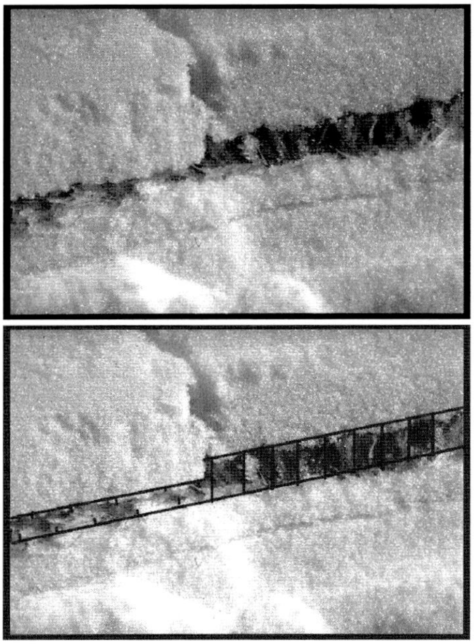

Figure 8.1 The upper photo shows a buried weak layer above a hoarfrost sheet. Photo from Jamieson and Schweizer [99]. In the lower picture, a schematic drawing of fibers representing the partly fractured weak layer is overlying the photo. The snow bonds are intact on the right, fractured on the left [97, 98].

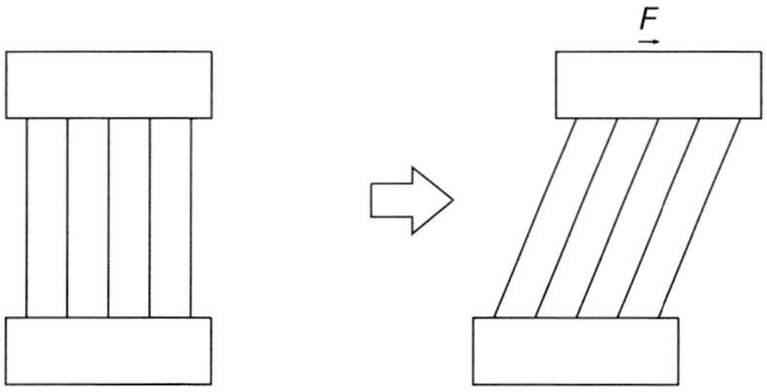

Figure 8.2 Schematic representation of the model under shear. The upper plate represents the snow slab, the lower plate the hoarfrost layer. A displacement Δx of the upper plate, orthogonal to the initial fiber direction, implies stretching of the fibers.

In this model, broken fibers may undergo a healing process. The ice grains are close to the melting temperature, and the broken fibers may easily resinter. At each time step, a probability p_s for starting such sintering is assumed. For a broken fiber to regain full strength, s number of time steps are necessary. The sintering probability p_s is assumed to be proportional to the square of the number of broken fibers, since two fiber ends are necessary for sintering.

In this model, a snow avalanche is predicted to start when the fracturing of bonds cannot be balanced by formation of new bonds.

Figure 8.3 Roots stretched across a tension crack. (Picture credit: belop GmbH, Switzerland.)

To back up the model, theoretical model results were tested by challenging experiments in a cold laboratory, using a force-controlled apparatus for slope angles between 0 and 60°. The experimental stress–strain curves thus obtained could be matched in good qualitative agreement with properly parameterized model results [97].

8.2 Shallow Landslides

There are different types of landslides and different landslide causes. We mention here briefly a few landslide models that involve fiber bundles.

Firstly, in a similar way as for slab snow avalanches in the preceding section, a buried weak layer may cause a shallow landslide to occur.

The weak layer is usually caused by infiltration of water, by rapid snow melting, or by heavy rainfall. This reduces the soil strength. The water-induced weakening is modeled by letting the strength distribution of the fibers in the fiber bundle depend on the water content. Modeling the time-dependent water infiltration for a given rainfall also allows estimation of the time to failure [100].

Secondly, roots have a stabilizing effect in soil and may inhibit landslides. Roots are very naturally modeled by fibers (see Figure 8.3). Cohen, Schwarz, and Or developed two fiber bundle models [101] for shallow landslides. The more complex of their models includes root–soil interfacial interactions and root geometrical properties. The models provide estimates for the maximum root reinforcement as a function of displacement, an important quantity in slope stability calculations of rooted soils. The model predictions were compared with experimental results for soil with spruce roots.

Thirdly, a more complex model of landslides has been put forward by Lehmann and Or [100]. They envisage the hill slope to consist of soil columns with hexagonal cross section (Figure 8.4). The slope of the hill is determined by the ratio of the average difference between two neighboring columns and the column diameter.

With complete parameterization, the authors were able to produce simulation results that gave size and frequency statistics obeying power laws with exponents similar to values from real landslide inventories. For details, we refer to the original publication [100]. Modeling the dynamics of the system has two basic ingredients. The first ingredient is similar to the sandpile cellular automaton of Bak *et al.* [102]. Each hexagonal column may store a prescribed amount of earth, and when this threshold capacity is reached, earth is distributed downhill to adjacent columns. As in the sandpile model, a small amount of material added to the system may induce an avalanche.

Figure 8.4 A landslide triggering model. The hill slope is viewed as a collection of soil columns with hexagonal cross sections. In analogy with the sandpile cellular automaton [102], each column can accommodate a certain amount of earth, in units pictured as spheres. Here, loads with more than four units induce a chain reaction moving material in the downslope direction. The stabilizing mechanism is modeled as intercolumn fiber bundles. (From Ref. [100].)

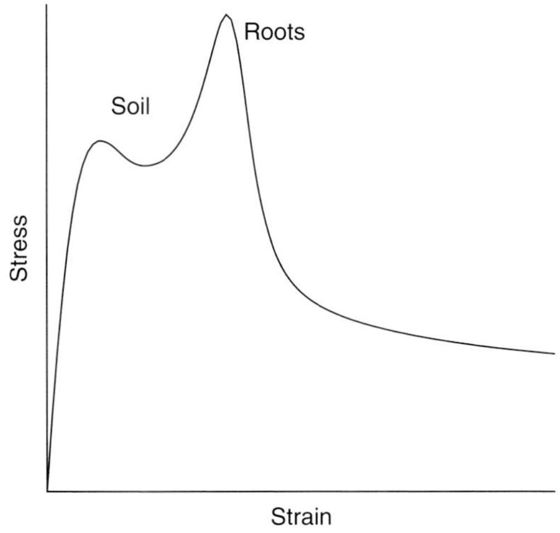

Figure 8.5 Stress–strain relation for a fiber mixture representing soil samples containing roots.

The second ingredient is that the columns are interconnected by frictional or mechanical bonds (like roots). These stabilizing forces are modeled by intercolumn fiber bundles (Figure 8.4). Note that the characteristic strain thresholds for basic soil strength and for roots are expected to be different. It is, therefore, natural to consider these fiber bundles as a mixture of two components, one representing soil and the other representing roots. The strain–stress relation will then be determined by the mixture formula (2.26). If the elastic constants for the two components are sufficiently different, one expects a stress–strain relation with two distinct peaks, as indicated in Figure 8.5. The graph in Figure 8.5 represents the experimental results of Fannin, Eliadorani, and Wilkinson reasonably well (see Lehmann and Or [100]).

The purpose of this chapter was to point to some unconventional applications of fiber bundle models. We expect the reader soon to encounter additional unorthodox and productive uses of fiber bundle models in other fields.

Appendix A
Mathematical Toolbox

A.1
Lagrange's Inversion Theorem

Lagrange's inversion theorem [103] can be stated as follows: let

$$x = a + t\phi(x) \tag{A.1}$$

define x as a function of t, and let $\phi(x)$ and $f(x)$ be analytic functions. For sufficiently small t, the following expansion is valid:

$$f(x) = f(a) + \sum_{n=1}^{\infty} \frac{t^n}{n!} \frac{\mathrm{d}^{n-1}}{\mathrm{d}a^{n-1}} \left[f'(a)\phi(a)^n\right]. \tag{A.2}$$

A useful special case is $\phi(x) = e^x$ and $f(x) = x$, for which (A.2) takes the form

$$x = a + \sum_{n=1}^{\infty} \frac{n^{n-1}}{n!} t^n e^{na}, \tag{A.3}$$

and, in particular, with $a = 0$:

$$\sum_{n=1}^{\infty} \frac{n^{n-1}}{n!} t^n = x. \tag{A.4}$$

By (A.1), it follows that this sum satisfies

$$xe^{-x} = t. \tag{A.5}$$

$t(x)$ has a maximum at $x = 1$, so that x is a double-valued function of t. Clearly, the branch with $0 \leq x \leq 1$ is the relevant one. The summation formula (A.4–A.5) has been used repeatedly in Chapter 3.

A proof of Lagrange's formula (A.2) using the Cauchy integral theorem runs as follows. Let C be a contour surrounding the point a in the complex z plane, and let t be so small that $|t\phi(z)| < |z - a|$ on and inside C. In the integral along C in the counterclockwise direction,

$$I = \frac{1}{2\pi i} \oint f(z) \frac{1 - t\phi'(z)}{z - a - t\phi(z)} \mathrm{d}z, \tag{A.6}$$

The Fiber Bundle Model: Modeling Failure in Materials, First Edition.
Alex Hansen, Per C. Hemmer and Srutarshi Pradhan.

we may, therefore, expand the integrand in powers of t:

$$\frac{1 - t\phi'(z)}{z - a - t\phi(z)} = \frac{1}{z-a} + \sum_{n=1}^{\infty} t^n \left[\frac{\phi(z)^n}{(z-a)^{n+1}} - \frac{\phi(z)^{n-1}\phi'(z)}{(z-a)^n} \right]$$

$$= \frac{1}{z-a} - \frac{\mathrm{d}}{\mathrm{d}z} \sum_{n=1}^{\infty} \frac{t^n \phi(z)^n}{n(z-a)^n} \tag{A.7}$$

Upon insertion into (A.6), the first term gives $f(a)$. In the second term, we perform a partial integration. Thus,

$$I = f(a) + \frac{1}{2\pi i} \sum_{n=1}^{\infty} \frac{t^n}{n} \oint \frac{f'(z)\phi(z)^n}{(z-a)^n} \,\mathrm{d}z. \tag{A.8}$$

By a Taylor expansion in $z - a$ of $f'(z)\phi(z)^n \equiv g(z)$,

$$g(z) = \sum_{k=0}^{\infty} \frac{1}{k!}(z-a)^k \frac{\mathrm{d}^k}{\mathrm{d}a^k} g(a), \tag{A.9}$$

only the term with $k = n - 1$ produces a simple pole in the integrand of (A.8). Consequently,

$$I = f(a) + \sum_{n=1}^{\infty} \frac{t^n}{n!} \frac{\mathrm{d}^{n-1}}{\mathrm{d}a^{n-1}} \left[f'(a)\phi(a)^n \right]. \tag{A.10}$$

On the other hand, we may calculate the integral (A.6) directly. It is of the form

$$I = \frac{1}{2\pi i} \oint f(z) \frac{F'(z)}{F(z)} \,\mathrm{d}z, \tag{A.11}$$

with $F(z) = z - a - t\phi(z) = z - x$ using (A.1). Thus, the integrand in (A.11) has a simple pole at $z = x$, with the residue $f(x)$. Hence,

$$I = f(x). \tag{A.12}$$

A combination of equations (A.12) and (A.10) produces Lagrange's theorem (A.2).

A.2 Some Theorems in Combinatorics

A.2.1 Basic Selection Theorems

(i) Out of N distinct elements, one can choose a group of n ordered elements in

$$N \cdot (N-1) \cdots (N-n+1) = \frac{N!}{(N-n)!} \tag{A.13}$$

different ways.

(ii) Out of N distinct elements, one can choose a group of n elements in

$$\binom{N}{n} = \frac{N}{n!(N-n)!} \tag{A.14}$$

different ways. Equation (A.14) defines a binomial coefficient.

(iii) n distinct elements are distributed randomly between N boxes. The probability that one specified box contains precisely k elements equals

$$\binom{n}{k}\left(\frac{1}{N}\right)^k\left(1-\frac{1}{N}\right)^{n-k} \tag{A.15}$$

A.2.2
A Distribution with Restrictions

Let n distinct elements be distributed among N numbered boxes, with the same *a priori* probability to find an element in any box. We seek the probability $p(n, N)$ that the last box (number N) contains at most zero elements, box number $N-1$ at most one element, and so on. In general, the last p boxes shall contain at most $p-1$ elements.

The probability that box number 1 contains precisely k elements equals

$$\binom{n}{k}\left(\frac{1}{N}\right)^k\left(1-\frac{1}{N}\right)^{n-k}, \tag{A.16}$$

by eq. (A.15) With k elements in box number 1, the probability that remaining $n-k$ elements are distributed in the remaining $N-1$ boxes, fulfilling the given restrictions, is $p(n-k, N-1)$. Thus,

$$p(n,N) = \sum_{k=0}^{n}\binom{n}{k}\left(\frac{1}{N}\right)^k\left(1-\frac{1}{N}\right)^{n-k} p(n-k,N-1). \tag{A.17}$$

We will prove by induction on N that the probability we seek is

$$p(n,N) = 1-\frac{n}{N}. \tag{A.18}$$

It is easy to check directly that (A.18) is valid for $N = 1$ and for $N = 2$. We assume that it is valid for $N-1$ and evaluate the right-hand side of (A.17).

$$\begin{aligned} p(n,N) &= \sum_{k=0}^{n}\binom{n}{k}\left(\frac{1}{N}\right)^k\left(1-\frac{1}{N}\right)^{n-k}\left(1-\frac{n-k}{N-1}\right) \\ &= 1-\sum_{k=0}^{n-1}\frac{n!}{k!(n-k)!}\left(\frac{1}{N}\right)^k\left(\frac{N-1}{N}\right)^{n-k}\frac{n-k}{N-1} \\ &= 1-\frac{n}{N}\sum_{k=0}^{n-1}\frac{(n-1)!}{k!(n-k-1)!}\left(\frac{1}{N}\right)^k\left(\frac{N-1}{N}\right)^{n-1-k} \\ &= 1-\frac{n}{N}, \end{aligned} \tag{A.19}$$

as the sum is the binomial expansion of $[1/N + (1-1/N)]^{n-1}$.

In conclusion, the validity of (A.18) for $N-1$ implies the validity of (A.18) for N, which completes the induction proof.

For the application in the Chapter 3, $n = N-1$, which gives

$$p(N-1, N) = \frac{1}{N}. \tag{A.20}$$

A.3
Biased Random Walks

A random walker on the integer sites of the z axis has the probability p to take the next step in the positive direction and the probability $q = 1-p$ to take the step in the negative direction. The step length is unity. When $p \neq q$, the walk is *biased*.

A.3.1
Probability of No Return

We start by determining the probability ξ to reach the neighboring site in the positive direction for the first time. Nothing is assumed about the number of steps used to accomplish this. (A general reference for random walk properties is Feller's book [27]).

With the starting point $z = 0$, ξ is the probability to reach $z = 1$. After the first step, the walker is at $z = 1$ with probability p or at $z = -1$ with probability q. Starting from $z = -1$, the probability to reach $z = 1$ equals ξ^2, so that the probability ξ to reach $z = 1$ for the first time satisfies

$$\xi = p + q\xi^2. \tag{A.21}$$

The relevant solution of (A.21) is

$$\xi = \frac{1-\sqrt{1-4pq}}{2q}. \tag{A.22}$$

The solution with the positive square root is unphysical, since it behaves like $1/q$ for small q.

Using $1-4pq = (p+q)^2 - 4pq = (p-q)^2$, we have

$$\xi = \frac{1-|p-q|}{2q} = \begin{cases} 1 & \text{for } p > q \\ p/q & \text{for } p < q. \end{cases} \tag{A.23}$$

Hence, the walker reaches $z = 1$ with probability 1 when the bias is favorable. When the bias is unfavorable, however, there is a nonzero probability $1 - p/q$ that the site $z = 1$ is never reached.

Now it is easy to find the probability $p(\text{no return})$ that the walker never returns to the starting point $z = 0$. We first assume $p < q$. With probability q the first step ends up at $z = -1$, but according to (A.23), at $z = -1$ with probability $1 - p/q$ the walker will never reach $z = 0$ later on. Thus,

$$p(\text{no return}) = q(1-p/q) = q - p(p < q). \tag{A.24}$$

The walker may, of course, take the first step to $z = 1$, but, according to (A.23), with probability 1 will reach $z = 0$ later on, so this does not increase the probability of no return.

Due to the left/right symmetry ($p \leftrightarrow q$), we have in general

$$p(\text{no return}) = |q - p|. \tag{A.25}$$

Conversely, the probability that the walker returns at least once to his starting point is

$$p(\text{return}) = 1 - |p - q|. \tag{A.26}$$

A.3.2
Gambler's Ruin

In the classical ruin problem, a gambler with an initial capital of z euros plays a series of independent games against a bank. In each game, the bank wins or loses one euro with probabilities p and q, respectively. It is assumed that $p > q$, the game favors the bank. The game continues until the gambler is ruined. Here we consider the case in which the bank has unlimited resources.

The ruin problem is straightforwardly formulated as a biased random walk problem of the type discussed earlier. The walk starts at the site z and continues until the site $z = 0$ is reached.

We seek that the probability $\eta(z, n)$ that the walk starting at z ends with the nth step at $z = 0$ (gambler's ruin in the nth game). After the first step, the walker's position is $z + 1$ or $z - 1$, so that

$$\eta(z, n + 1) = q\ \eta(z + 1, n) + p\ \eta(z - 1, n). \tag{A.27}$$

This is also valid for $z = 1$ when we define $\eta(0, n) = 0$ for $n > 0$, and for $n = 0$, when we define $\eta(z, 0) = 0$ for $z > 0$ while $\eta(0, 0) = 1$.

We obtain a relation for the generating function

$$G(z, s) = \sum_{n=0}^{\infty} \eta(z, n)s^n \tag{A.28}$$

by multiplying (A.27) by s^{n+1} and summing over n. We obtain

$$G(z, s) = qsG(z + 1, s) + ps\ G(z - 1, s). \tag{A.29}$$

For $z = 0$, we must have

$$G(0, s) = 1, \tag{A.30}$$

in view of the definitions $\eta(0, n) = 0$ for $n > 0$ and $\eta(0, 0) = 1$ made above.

The difference equation (A.29) allows solutions of the form $G(z, s) = \zeta(s)^z$. Insertion yields

$$1 = qs\zeta + ps\zeta^{-1}, \quad \text{i.e.}\ \zeta_{\pm} = \frac{1 \pm \sqrt{1 - 4pqs^2}}{2qs}. \tag{A.31}$$

The solution with the plus sign is unphysical since ζ_+^z diverges for $s \to 0$. Thus, the generating function is

$$G(z,s) = \left[\frac{1-\sqrt{1-4pqs^2}}{2qs}\right]^z. \tag{A.32}$$

Note that the boundary condition (A.30) is fulfilled.

We may obtain the ruin probabilities $\eta(z,n)$ by expanding the generating function in powers of s, see eq. (A.28). Of special interest in Chapter 3 is $\eta(1,n)$. Expanding $\sqrt{1-4pqs^2}$ according to the binomial theorem, we obtain

$$G(1,s) = -\frac{1}{2q}\sum_{k=1}^{\infty}\binom{1/2}{k}(-4pq)^k\, s^{2k-1}. \tag{A.33}$$

Hence

$$\eta(1,2k-1) = -\frac{(-4pq)^k}{2q}\binom{1/2}{k}. \tag{A.34}$$

The binomial coefficient may be written

$$\binom{\frac{1}{2}}{k} = \frac{\frac{1}{2}\cdot\left(-\frac{1}{2}\right)\cdot\left(-\frac{3}{2}\right)\cdots\left(-k+\frac{3}{2}\right)}{k!} = \frac{(-1)^{k-1}\,(2k-2)!}{k!(k-1)!2^{2k-1}}. \tag{A.35}$$

Consequently we obtain

$$\eta(1,2k-1) = \frac{(pq)^k\,(2k-2)!}{q\;k!(k-1)!}, \tag{A.36}$$

i.e.

$$\eta(1,n) = \frac{(n-1)!}{\frac{n-1}{2}!\,\frac{n+1}{2}!}\,(pq)^{n/2}\,\sqrt{p/q} = \frac{1}{n}\binom{n}{\frac{n-1}{2}}(pq)^{n/2}\,\sqrt{p/q}. \tag{A.37}$$

We note in passing, without providing proof, the probability of ruin after n games for a gambler with an initial capital z:

$$\eta(z,n) = \frac{z}{n}\binom{n}{\frac{n-z}{2}}(pq)^{n/2}\,(p/q)^{z/2}. \tag{A.38}$$

Of greater interest in our context is large n, that is, the probability that the gambler avoids ruin for quite a while, or, alternatively, that the walker makes a very long detour before hitting the neighboring site $z = 0$. By using the Stirling approximation for the factorial, $n! \simeq (n/e)^n\sqrt{2\pi n}$, and the limit $(1 + a/n)^n \to e^a$, we obtain

$$\eta(1,n) \simeq \sqrt{2/\pi}\; n^{-3/2}\,(4pq)^{n/2}\sqrt{p/q}\ \text{(large } n\text{)}. \tag{A.39}$$

This result is used in Section 3.2.2.

A.4
An Asymmetrical Unbiased Random Walk

The exact random walk corresponding to the force steps $\Delta F \equiv f$ in a fiber bundle at criticality has the following step probability density $\rho(f)$

$$\rho(f) = \begin{cases} e^{-1-f} & \text{for } f \geq -1 \\ 0 & \text{for } f < -1. \end{cases} \tag{A.40}$$

This follows from formula (3.98) and the criticality property (2.14), $1 - P(x_c) = x_c\, p(x_c)$. The scale factor x_c has for convenience been set to unity.

The random walk paths of interest for fiber bundle bursts of size Δ are of the following type: on each of the first $\Delta - 1$ steps, the position of the walker is to the left of the starting point, but after step no. Δ, the position exceeds the starting point. The probability of such processes equals

$$\text{Prob}(\Delta) = \int_{-1}^{0} \rho(f_1)\, \mathrm{d}f_1 \int_{-1}^{-f_1} \rho(f_2)\, \mathrm{d}f_2 \int_{-1}^{-f_1-f_2} \rho(f_3)\, \mathrm{d}f_3 \ldots$$
$$\int_{-1}^{-f_1-f_2\ldots-f_{\Delta-2}} \rho(f_{\Delta_1})\, \mathrm{d}f_{\Delta-1} \int_{-f_1-f_2\ldots-f_{\Delta-1}}^{\infty} \rho(f_\Delta)\, \mathrm{d}f_\Delta . \tag{A.41}$$

The last exponential integral gives a factor $\exp\,(-1+f_1+f_2+\ \ldots f_{\Delta-1})$, and since $\rho(f)e^f = e^{-1}$, we end up with

$$\text{Prob}(\Delta) = e^{-\Delta} \int_{-1}^{0} \mathrm{d}f_1 \int_{-1}^{-f_1} \mathrm{d}f_2 \int_{-1}^{-f_1-f_2} \mathrm{d}f_3 \ldots \int_{-1}^{-f_1-f_2\ldots-f_{\Delta-2}} \mathrm{d}f_{\Delta-1}. \tag{A.42}$$

In terms of the new variables

$$y_1 = -f_1, y_2 = -f_1 - f_2, \ldots, y_{\Delta-1} = -f_1 - f_2 \ldots - f_{\Delta-1}, \tag{A.43}$$

we obtain

$$\text{Prob}(\Delta) = e^{-\Delta} \int_{0}^{1} \mathrm{d}y_1 \int_{0}^{1+y_1} \mathrm{d}y_2 \int_{0}^{1+y_2} \mathrm{d}y_3 \ldots \int_{0}^{1+y_\Delta} \mathrm{d}y_{\Delta-1}. \tag{A.44}$$

Defining $V_d(y) = 1$ and the iteration

$$V_d(y) = \int_{0}^{1+y} V_{d-1}(z)\, \mathrm{d}z, \tag{A.45}$$

we have

$$\text{Prob}(\Delta) = e^{-\Delta}\, V_{\Delta-1}(0). \tag{A.46}$$

The first values are $V_1(y) = y + 1$ and $V_2(y) = (y^2 + 4y + 3)/2$. We now solve the iteration (A.45) by induction. The induction assumption

$$V_{d-1}(y) = \frac{1}{d!} \sum_{i=0}^{d-1} d^{d-i-1} \binom{d-1}{i} (i+1)y^i \tag{A.47}$$

is easily verified to be true for $d = 2$ and $d = 3$. We suppose (A.47) to be true up to some value of $d - 1$, and use (A.45) to compute V_d. The integration in (A.45) is trivial, leaving

$$V_d(y) = \sum_{i=0}^{d-1} \binom{d-1}{i} d^{d-i-1} (1+y)^{i+1}. \tag{A.48}$$

We now convert this expression to a power series in y,

$$V_d(y) = \frac{1}{d!} \sum_{m=0} S(m)\, y^m. \tag{A.49}$$

Using in equation (A.48) the binomial theorem to expand $(1+y)^{i+1}$, we obtain

$$\begin{aligned} S(m) &= \sum_{i=m-1}^{d-1} \binom{d-1}{i} \binom{i+1}{m} d^{d-i-1} \\ &= \sum_{i=m-1}^{d-1} \binom{d-1}{i} \left[\binom{i}{m} + \binom{i}{m-1} \right] d^{d-i-1}. \end{aligned} \tag{A.50}$$

We have used the identity

$$\binom{i+1}{m} = \binom{i}{m} + \binom{i}{m-1}, \tag{A.51}$$

which is easily proved using the definition (A.14) of binomial coefficients.
To evaluate $S(m)$, we differentiate the binomial expression

$$\sum_{i=0}^{d-1} \binom{d-1}{i} x^i = (1+x)^{d-1} \tag{A.52}$$

m times with respect to x:

$$\sum_{i=0}^{d-1} \binom{d-1}{i} i(i-1) \ldots (i-m+1)x^{i-m} = (d-1)(d-2) \ldots (d-m)(1+x)^{d-1-m}. \tag{A.53}$$

Division by $m!$ yields

$$\sum_{i=0}^{d-1} \binom{d-1}{i} \binom{i}{m} x^{i-m} = \frac{(d-1)!}{m!(d-1-m)!}(1+x)^{d-1-m}. \tag{A.54}$$

Putting now $x = 1/d$, and multiplying both sides by d^{d-1-m}, we obtain the first term in (A.50):

$$S_1(m) = \sum_{i=0}^{d-1} \binom{d-1}{i} \binom{i}{m} d^{d-i-1} = \frac{(d-1)!}{m!(d-1-m)!}(1+d)^{d-1-m}. \tag{A.55}$$

The second sum in (A.50) is the same as the first one with m replaced by $m - 1$. Hence,

$$S(m) = S_1(m) + S_1(m-1) = \binom{d}{m} (1+d)^{d-m} \frac{1+m}{1+d}. \tag{A.56}$$

Finally Eq. (A.49) gives

$$V_d(y) = \frac{1}{(1+d)!} \sum_{m=0}^{d} \binom{d}{m} (d+1)^{d-m}(1+m)\, y^m. \tag{A.57}$$

Since this is in accordance with the induction assumption (A.47), and since (A.47) is correct for small d, the induction proof is completed.

By (A.46) the probability we seek is

$$\text{Prob}(\Delta) = e^{-V}\; V_{\Delta-1}(0) = \frac{e^{-\Delta}\; \Delta^{\Delta-1}}{\Delta!}. \tag{A.58}$$

This result is used in Section 6.1.1.1.

A.5
Brownian Motion as a Scaled Random Walk

Earlier we considered random walks (with unit step length) on integers. Assuming the walks to be unbiased and that the walker starts at the origin, let us now scale lengths so that each step has a much shorter length ξ and that the time between two consecutive steps is τ. During the time t, the walker performs t/τ steps and the displacement is x/ξ, with x a multiple of ξ. Only multiples of ξ and τ occurs in the problem, but in the limit $\xi \to 0, \tau \to 0$ *any* displacement x and *all* times t become possible.

The total displacement during time t is the sum of t/τ independent steps each having zero mean and a variance ξ^2. The variance of the total displacement during the time t is therefore $\xi^2 t/\tau$. To obtain sensible results in the limit, the variance must be finite when ξ and τ approach zero. Thus, we require

$$\frac{\xi^2}{\tau} = \text{constant} = 1, \tag{A.59}$$

Where, for simplicity, the constant is chosen to be unity. The variance of the total displacement is then t.

Since the total displacement x consists of many independent and identically distributed steps, the central limit theorem assures us that x is normally distributed:

$$\rho(x,t) = \frac{1}{\sqrt{2\pi t}}\; e^{-x^2/2t}. \tag{A.60}$$

The resulting continuous process is a Wiener process or *Brownian motion*. The irregular motion of pollen particles in a liquid, observed by Robert Brown, is, as in our case, the result of many minute displacements in rapid succession.

A standard notation of the Brownian process is to use $W(t)$ as the position variable, with $W(0) = 0$. Its derivation as the limit of a random walk makes it clear that the following properties hold:

$$\langle W(t) \rangle = 0 \tag{A.61}$$

$$\langle W(t)^2 \rangle = t \tag{A.62}$$

$$W(s+t) - W(s) \text{ is independent of } W(\tau) \text{ for } \tau < s \tag{A.63}$$

$$\text{Prob}(W(t) \in [x_1, x_2]) = \int_{x_1}^{x_2} \rho(x,t), \text{ with } \rho(x,t) \text{ given in (A.60).} \tag{A.64}$$

Upon scaling time with a factor c, the Brownian motion scales as

$$c^{-1/2} W(ct) = W(t). \tag{A.65}$$

This is valid since both sides of (A.65) have the same (zero) average and by property (A.62) the same variance.

Appendix B
Statistical Toolbox

B.1
Stochastic Variables, Statistical Distributions

If x is a continuous stochastic variable, there will be a probability density $p(x)$ associated with it. That is, there is a probability $p(x)dx$ that x will have a value within dx of x. The cumulative probability is defined as

$$P(x) = \int_{-\infty}^{x} p(x')\,\mathrm{d}x'. \tag{B.1}$$

This quantity gives the probability that the stochastic variable to have value x or less. If x is limited to the interval $x_< \leq x \leq x_>$, we have $P(x \leq x_<) = 0$ and $P(x \geq x_>) = 1$.

B.1.1
Change of Variable

We have our stochastic variable x and its probability density $p(x)$. We define a new variable y through the relation $x = x(y)$. The variable y will have a probability density $q(y)$ given by

$$p(x)\,\mathrm{d}x = p(x(y))\left(\frac{\mathrm{d}x}{\mathrm{d}y}\right)\,\mathrm{d}y = q(y)\,\mathrm{d}y, \tag{B.2}$$

so that $q(y) = p(x(y))(\mathrm{d}x(y)/\mathrm{d}y)$.

In terms of the cumulative probabilities, we then have the relation

$$Q(y) = \int_{-\infty}^{y} q(y')\,\mathrm{d}y' = \int_{-\infty}^{y} p(x(y'))\left(\frac{\mathrm{d}x(y')}{\mathrm{d}y'}\right)\,\mathrm{d}y' = \int_{-\infty}^{x(y)} p(x')\,\mathrm{d}x' = P(x(y)). \tag{B.3}$$

B.1.1.1 A Useful Interpretation

It is very useful to think of the relation between x and P as a *change of variable,* that is,

$$x \to P(x), \tag{B.4}$$

The Fiber Bundle Model: Modeling Failure in Materials, First Edition.
Alex Hansen, Per C. Hemmer and Srutarshi Pradhan.

so that P is a variable in the same way that x is a variable. We have that

$$\mathrm{d}P = 1 \cdot \mathrm{d}P = \left(\frac{\mathrm{d}P(x)}{\mathrm{d}x}\right) \mathrm{d}x = p(x)\,\mathrm{d}x, \tag{B.5}$$

This shows that the probability density for the stochastic variable P is the *uniform distribution on the unit interval.* As we will see in the following, it is in many instances very convenient to work with the variable P and then perform the inverse transformation $x = x(P)$.

B.1.2
The Characteristic Function

Given a stochastic variable and its associated probability density function $p(x)$, we may define the characteristic function

$$\phi_x(y) = \langle e^{iyx}\rangle = \int_{-\infty}^{+\infty} \mathrm{d}x\, e^{ixy} p(x), \tag{B.6}$$

which is the Fourier transform of the probability density $p(x)$. The normalization of the probability density function demands that

$$\phi(0) = \int_{-\infty}^{+\infty} \mathrm{d}x\, p(x) = 1. \tag{B.7}$$

The characteristic function furthermore obeys the inequality

$$|\phi(y)| = \left|\int_{-\infty}^{+\infty} \mathrm{d}x\, e^{ixy} p(x)\right| \leq \int_{-\infty}^{+\infty} \mathrm{d}x\, |e^{ixy} p(x)| = \int_{-\infty}^{+\infty} \mathrm{d}x\, p(x) = 1. \tag{B.8}$$

The nth moment of the probability density $p(x)$ is defined as

$$\langle x^n\rangle = \int_{-\infty}^{+\infty} \mathrm{d}x\, x^n p(x). \tag{B.9}$$

We expand the characteristic function in a series in the variable y,

$$\phi(y) = \sum_{n=0}^{\infty} \left(\frac{(iy)^n}{n!}\right) \langle x^n\rangle. \tag{B.10}$$

Hence, the moments are generated by the characteristic function,

$$\langle x^n\rangle = (-i)^n \left(\frac{\mathrm{d}^n \phi(y)}{\mathrm{d}y^n}\right)_{y=0}. \tag{B.11}$$

If we expand the logarithm of the characteristic function in a series in y, we find

$$\ln[\phi(y)] = \sum_{n=0}^{\infty} \left(\frac{(iy)^n}{n!}\right) \kappa_n, \tag{B.12}$$

where

$$\kappa_n = (-i)^n \left(\frac{\mathrm{d}^n \ln[\phi(y)]}{\mathrm{d}y^n}\right)_{y=0} \tag{B.13}$$

are the *cumulants*. The three first cumulants are

$$\kappa_1 = \langle x \rangle, \tag{B.14}$$

$$\kappa_2 = \langle x^2 \rangle - \langle x \rangle^2, \tag{B.15}$$

and

$$\kappa_3 = \langle x^3 \rangle - 3\langle x^2 \rangle \langle x \rangle + 2\langle x \rangle^3. \tag{B.16}$$

If the underlying distribution $p(x)$ is Gaussian, then

$$\kappa_n = 0 \quad \text{for} \quad n \geq 3. \tag{B.17}$$

B.1.3
The Central Limit Theorem

The Gaussian distribution is ubiquitous. Why? This is due to the *Central limit theorem*, which we will now derive [104].

We assume that we have N independent elements $x_j, j = 1, \ldots N$ drawn from the same probability distribution $p(x)$. We assume that $\langle x \rangle = 0$ and that all moments $\langle x^n \rangle$ are finite. The characteristic function associated with $p(x)$ is $\phi(y)$.

We now consider the sum of the variables x_j weighted by [1] $N^{-1/2}$

$$X = \frac{1}{N^{1/2}} \sum_{j=1}^{N} x_j. \tag{B.18}$$

The characteristic function of this variable, X, is

$$\phi_X(y) = \langle e^{iyX} \rangle = \langle e^{(iyN^{-1/2}) \sum_{j=1}^{N} x_j} \rangle = \left[\phi \left(\frac{y}{N^{1/2}} \right) \right]^N. \tag{B.19}$$

Since the moments of $p(x)$ all exist, we may expand $\phi(y/N^{1/2})$ in a Taylor series, see Eq. (B.10),

$$\phi \left(\frac{y}{N^{1/2}} \right) = 1 - \left(\frac{y^2}{2N} \right) \langle x^2 \rangle + \mathcal{O} \left(\frac{y^3}{N^{3/2}} \right). \tag{B.20}$$

As N increases, the higher order terms vanish compared to the second term. Hence, if we combine this equation with Eq. (B.19), we find

$$\phi_X(y) = \lim_{N \to \infty} \left[1 - \frac{y^2 \langle x^2 \rangle}{2N} \right]^N = e^{-y^2 \langle x^2 \rangle / 2}. \tag{B.21}$$

We Fourier-transform this expression from the variable y to the variable X and find

$$p_X(X) = \frac{e^{-X^2/2\langle x^2 \rangle}}{\sqrt{2\pi \langle x^2 \rangle}}, \tag{B.22}$$

1) Hence, $X/N^{1/2} \to \langle x \rangle = 0$ in the limit $N \to \infty$.

which is the Gaussian distribution.

The assumptions we made were that (1) the first moment $\langle x \rangle = 0$ and (2) all moments $\langle x^n \rangle$ are finite. The first assumption may always be fulfilled by shifting the origin of the x axis. Hence, we are left with the second assumption only. The distribution of sums – that is, averages – over variables for which all statistical moments are finite, will always tend toward the Gaussian distribution.

B.2 Order Statistics

Order statistics plays an important role in the analysis of fiber bundle model. It is not a well-known subject, but it is very useful in the study of fiber bundle models. Hence, we provide the reader here with a short review of some of its main concepts and results.

B.2.1 Ordering the Variables

Suppose now that we do not draw a value for the stochastic variable x once, but N times. That is, we generate a sequence x_1 to x_N of values, all distributed according to the probability density $p(x)$. What can we say about this sequence? This is the subject of *order statistics* [105].

We rearrange the sequence $x_1, \cdots, x_N$ in an ascending order such that $x_1 \leq x_2 \leq \cdots \leq x_N$.

Let us now generate another sequence of N values $x_1 \leq \cdots x_N$. To distinguish this second sequence from the first, we introduce the notation $x_1(i) \leq \cdots x_N(i)$, where $i = 1$ or 2. We do, however, not stop here. We generate K sequences so that $i = 1, \cdots, K$. This gives us a matrix of values

$$\begin{gathered} x_1(1) \leq x_2(1) \leq x_3(1) \leq \cdots \leq x_{N-2}(1) \leq x_{N-1}(1) \leq x_N(1) , \\ x_1(2) \leq x_2(2) \leq x_3(2) \leq \cdots \leq x_{N-2}(2) \leq x_{N-1}(2) \leq x_N(2) , \\ \vdots \quad \vdots \quad \vdots \\ x_1(K) \leq x_2(K) \leq x_3(K) \leq \cdots \leq x_{N-2}(K) \leq x_{N-1}(K) \leq x_N(K) . \end{gathered} \tag{B.23}$$

Let us now concentrate on one column of this matrix,

$$\begin{aligned} &x_m(1) \ , \\ &x_m(2) \ , \\ &\vdots \\ &x_m(K) \ . \end{aligned} \tag{B.24}$$

Whereas the members of the sequence $x_1(k), \cdots x_N(k)$ are distributed according to the probability distribution $p(x)$, the sequence $x_m(1), \cdots x_m(K)$ is *not!* Why? It is because we have added the constraint that each member of the latter sequence is the mth smallest value of N samplings of the original distribution $p(x)$. The sequence is drawn from the distribution

$$p_N^m(x) = \frac{N!}{(N-m)!(m-1)!} P(x)^{m-1} (1 - P(x))^{N-m} p(x). \tag{B.25}$$

B.2.2
The Average of the mth Ordered Variable

What is the average of x_m? In order to calculate this, let us change our variable from x to P as described in Section B.1.1.1. In the variable P, the probability density p_N^m (B.25) becomes

$$p_N^m(P) = \frac{N!}{(N-m)!(m-1)!} P^{m-1} (1 - P)^{N-m}, \tag{B.26}$$

and we find easily that

$$\langle P \rangle = \int_0^1 \mathrm{d}P\, P\, p_N^m(P) = \frac{m}{N+1}. \tag{B.27}$$

Returning to the original variable x, we then find

$$\langle P(x_m) \rangle = \frac{m}{N+1}. \tag{B.28}$$

This is a very useful result. [2)]

B.3
The Joint Probability Distribution

The probability density to find a value x_m for the mth smallest variable and x_l for the lth smallest variable is [32]

$$\phi_N(x_m, x_l) = C P(x_m)^{m-1} [P(x_l) - P(x_m)]^{l-m-1} [1 - P(x_l)]^{N-l} p(x_m) p(x_l), \tag{B.29}$$

when $x_m < x_l$. The combinatorial factor is

$$C = \frac{N!}{(m-1)!(l-m-1)!(N-l)!}. \tag{B.30}$$

There are $m - 1$ values for x that are smaller than x_m. The probability for this is $P(x_m)^{m-1}$. There are $N - l$ values for x that are larger than l. The probability for this is $[1 - P(x)]^{N-l}$. Lastly, there are $l - m - 1$ values for x between x_m and x_l. This has probability $[P(x_l) - P(x_m)]^{l-m-1}$. The probability that these constraints are simultaneously fulfilled is the product of the three probabilities times the possible permutations between the values, given by the coefficient (B.30).

2) Equation (B.28) does not give $\langle x_m \rangle$ directly. However, in the large-N limit, $\langle P(x_m) \rangle$ will approach $P(\langle x_m \rangle)$.

B.3.1
Extreme Statistics

Extreme statistics is the part of order statistics where $m = 1$ or N, the smallest or the largest element [32]. Extreme statistics answers questions such as what is the largest flood to be expected over a 100-year period, what is the strength of a chain, and what is the expected lifetime of a device consisting of a large number of components that each may fail. In other words, it is a very useful discipline.

According to (B.25), the smallest element follows the distribution

$$p_N^1(x) = N(1 - P(x))^{N-1} p(x). \tag{B.31}$$

The corresponding cumulative probability is then

$$P_N^1(x) = 1 - (1 - P(x))^N. \tag{B.32}$$

In the same way, the largest element follows the distribution

$$p_N^N(x) = N P(x)^{N-1} p(x), \tag{B.33}$$

and the corresponding cumulative probability is

$$P_N^N(x) = P(x)^N. \tag{B.34}$$

B.3.2
The Largest Element: The Three Asymptotic Distributions

We will now discuss $P_N^N(x)$ and $P_N^1(x)$ in the limit when $N \to \infty$. As we shall see, there are only three different asymptotic extreme value distributions. Which of the three different asymptotic distributions that one finds depends on the initial probability density $p(x)$.

In Table (B.23), we listed K sequences of length N of independent samplings of the distribution $p(x)$. The largest member of each of these K sequences of length N, $x_N(1), x_N(2), \ldots\ x_N(K)$, follows the cumulative probability $P_N^N(x)$.

Among these K elements, each of which is the largest in its sequence of N samplings, there must be a largest one.

This largest element may also be seen as the largest among one single sequence of length KN. Hence, the distribution that this largest element follows must be $p_{KN}^{KN}(x)$.

For asymptotically large N, the functional form of $P_N^N(x)$ must be the same as the functional form of $P_{KN}^{KN}(x)$. Hence, they must be related through the equation

$$P_{KN}^{KN}(x) = P_N^N(a_K x + b_K), \tag{B.35}$$

since rescaling and shift of the variable x do not change the shape of the cumulative probability function when $P_N^N(x) \to P_{KN}^{KN}(x)$. Equation (B.35) is the Fréchet, Fisher, and Tippett stability criterion [32]. This is an equation that will give us a_K, b_K and the functional form of $P_N^N(x)$.

In order to simplify the notation, let us introduce $P_>(x)$ as the asymptotic probability as $N \to \infty$. Hence, combining the stability criterion with Eq. (B.34) we have

$$[P_>(x)]^K = P_>(a_K x + b_K). \tag{B.36}$$

B.3.2.1 The Gumbel Distribution

Let us now assume that

$$a_N = 1 \tag{B.37}$$

in the following. Equation (B.36) then becomes

$$[P_>(x)]^K = P_>(x + b_K). \tag{B.38}$$

Following Gumbel [32], we take the logarithm twice and get

$$\ln[-\ln[P_>(x)]] + \ln[K] = \ln[-\ln[P_>(x + b_K)]]. \tag{B.39}$$

This equation implies

$$\ln[-\ln[P_>(x)]] - \frac{x}{b_K}\ln[K] = C, \tag{B.40}$$

where C is a constant. To see this, calculate

$$\ln[-\ln[P_>(x + b_K)]] - \frac{x + b_K}{b_K}\ln[K] = \ln[-\ln[P_>(x)]] - \frac{x}{b_K}\ln[K] = C, \tag{B.41}$$

using Eq. (B.39).

We have from Eq. (B.36) that

$$[P_>(x)]^{KM} = P_>(x + b_{KM}) = [P_>(x + b_M)]^K = P_>(x + b_K + b_M). \tag{B.42}$$

Hence, we have

$$b_{KM} = b_K + b_M, \tag{B.43}$$

implying

$$b_K = -\frac{\ln[K]}{\alpha}, \tag{B.44}$$

which combined with Eq. (B.40) gives

$$\ln[-\ln[P_>(x)]] = \alpha(x - u), \tag{B.45}$$

where we have defined $u = -C/\alpha$. We solve this equation to find

$$P_>(x) = \exp[-e^{-\alpha(x-u)}], \tag{B.46}$$

which is the *Gumbel* cumulative probability. It is also called the *first asymptote.*

The Gumbel extreme value distribution will ensue if the underlying probability density $p(x)$ falls off faster than any power law as $x \to \infty$.

B.3.2.2 The Second and Third Asymptotes

Let us now assume

$$a_N \neq 1 \tag{B.47}$$

in Eq. (B.36).

Equation (B.36) implies that there is a value of $x = x'$ where the left- and Right-hand sides are equal. Hence, we must have that

$$x' = a_K x' + b_K. \tag{B.48}$$

This implies that at $x = x'$ we must have that

$$[P_>(x')]^K = P_>(x') \tag{B.49}$$

from Eq. (B.36). Equation (B.49) can only be satisfied if

$$P_>(x') = 0, \tag{B.50}$$

or

$$P_>(x') = 1. \tag{B.51}$$

In the case of (B.50), there is a smallest value $x_< = x'$ such that $P_>(x) = 0$ for $x \leq x_<$. for (B.51), there is a largest value $x_> = x'$ for which $P_>(x) = 1$ if $x \geq x_>$. Let us now shift the origin of x such that $x' = 0$. This implies that $b_K = 0$ in the following.

By repeating the logic behind Eq. (B.36), we find

$$[P_>(x)]^{KM} = P_>(a_{KM}x) = [P_>(a_M x)]^K = P_>(a_K \cdot a_M x). \tag{B.52}$$

Hence,

$$a_{KM} = a_K \cdot a_M. \tag{B.53}$$

This equation has as a solution

$$a_K = K^l, \tag{B.54}$$

where l is a constant.

We now take the logarithm twice in Eq. (B.36) to find

$$\ln[-\ln[P_>(x)]] + \ln[K] = \ln[-\ln[P_>(K^l x)]]. \tag{B.55}$$

This implies

$$\ln[-\ln[P_>(x)]] = \frac{\ln x}{l} - \frac{\ln[\nu]}{l}, \tag{B.56}$$

where ν is a constant. We solve this equation with respect to $P_>(x)$ to find

$$P_>(x) = e^{-(x/\nu)^{1/l}}. \tag{B.57}$$

Equation (B.57) may be written

$$-\ln\ [P_>(x)] = \left(\frac{x}{\nu}\right)^{1/l}. \tag{B.58}$$

When x is increased in this equation, the left-hand side decreases toward zero. If we now assume Eq. (B.50) is in effect, then $x \geq 0$ and hence $v > 0$ and $l < 0$. We set $k = -1/l\text{r}$, and Eq. (B.57) becomes

$$P_>(x) = e^{-(v/x)^k}. \tag{B.59}$$

This extreme value cumulative probability is usually referred to as the *second asymptote.* If the underlying probability distribution falls of as a power law as $x \to \infty$, this will be the one that we find for the extreme values.

If we now assume Eq. (B.51) is in effect, then $x < 0$. We then have to rewrite Eq. (B.56) as

$$\ln\,[-\ln\,[P_>(x)]] = \frac{\ln(x/v)}{l}, \tag{B.60}$$

where both x and v are less than zero. Hence, by Eq. (B.58), we must have that $k = 1/l > 0$ and we find

$$P_>(x) = e^{-(x/v)^k}. \tag{B.61}$$

This is the *Weibull cumulative probability.* The name *third asymptote* is also used.

The Weibull extreme distribution is found if the underlying probability density $p(x)$ is bounded from above, that is, there is a largest finite $x_>$.

B.3.3
The Smallest Element: The Weibull Distribution

In the context of failure threshold distributions, $p(x)$ must be bounded from below by an $x_> \geq 0$. The asymptotic distribution $P_N^1(x)$, defined in Eq. (B.32) – or in the notation introduced in Section B.3.2 – $P_<(x)$ may be found by inverting the x axis, that is, let $x \to -x$. With respect to the cumulative probabilities, this implies that

$$P_<(x) = 1 - P_>(-x). \tag{B.62}$$

Hence, the cumulative probability for the smallest value when the underlying probability distribution is bounded from below by a value $x_<$ is

$$P_<(x) = 1 - e^{-[(x-x_<)/(v-x_<)]^k}, \tag{B.63}$$

which is the *Weibull* cumulative probability. By calculating the average of this probability, we find that

$$\langle x_1 \rangle = v\Gamma\left[1 + \frac{1}{k}\right]. \tag{B.64}$$

Hence, the Weibull distribution has more to it than an "educated guess." It is the appropriate distribution for the weakest element of any distribution that is bounded from below. It is indeed the correct distribution for the strength of a chain when this depends on its weakest link.

Appendix C
Computational Toolbox

Even though there are many aspects of the fiber bundle model, which are analytically tractable, computer has become an indispensable tool for those who are studying it.

In this Appendix, we discuss two topics related to the implementation of fiber bundle models on computer. The first topic, generating random numbers, for example, for failure threshold distributions, concerns all implementations of fiber bundle models. The second topic, Fourier acceleration, is aimed at the implementation of the soft clamp fiber bundle model, described in Section 4.5. The calculation of the force distribution among the fibers in this model involves inverting large sparse matrices. This can be computationally very demanding. The Fourier acceleration technique has proved to be a very efficient method for speeding up this inversion.

C.1
Generating Random Numbers Following a Specified Probability Distribution

Here we explain how to generate sequences of numbers x that are distributed according to a given probability distribution $p(x)$. This is central to computational investigations of the fiber bundle model.

C.1.1
When the Cumulative Probability May Be Inverted

From the probability distribution $p(x)$, we get the cumulative probability

$$P(x) = \int_{-\infty}^{x} dx' \, p(x'). \quad (C.1)$$

In Section B.1.1, we pointed out that $x \to P(x)$ may be seen as a coordinate transformation from the variable x to the variable P. P is then uniformly distributed on the unit interval. Hence, if we use a random number generator to produce a random number r, which is uniformly distributed on the unit interval, we may simply set $P = r$ and invert it to get

$$x = P_{-1}(r). \quad (C.2)$$

The Fiber Bundle Model: Modeling Failure in Materials, First Edition.
Alex Hansen, Per C. Hemmer and Srutarshi Pradhan.

As an example, consider the distribution $p(x) = x^{-1/2}/2$ on the unit interval. The corresponding cumulative distribution is $P(x) = x^{1/2}$. If we generate random number r, $x = r^2$ will then be distributed according to $p(x) = x^{-1/2}/2$.

As a second example, we may take the distribution $p(x) = \exp(-x)$ where $0 \le x < \infty$. The cumulative probability is $P(x) = 1 - \exp(-x)$. With a random number r, we then have $x = -\ln(r)$. [1)]

C.1.2
Gaussian Numbers

The Gaussian distribution is so common that there are very efficient algorithms to generate numbers following it, for example, the *Box-Müller algorithm* [106].

The Gaussian distribution is

$$p(x) = \frac{e^{-x^2/2}}{\sqrt{2\pi}}, \tag{C.3}$$

where $-\infty < x < +\infty$. We have here rescaled the variable x so that $\langle x^2 \rangle = 1$. The trouble is that the cumulative probability $P(x) = [1 + \mathrm{erf}(x/\sqrt{2})]/2$ cannot be inverted with respect to x.

However, if we consider a pair of number (x,y), both distributed according to (C.3), their joint probability distribution will be $p(x, y) = p(x)p(y)$. If we now interpret (x,y) as a point $\vec{R}$ in a plane, we may convert into polar coordinates (R, ϕ). In these coordinates, the joint probability distribution becomes

$$p(R, \phi) = \frac{Re^{-R^2/2}}{2\pi}, \tag{C.4}$$

where $0 \le R < \infty$ and $0 \le \phi \le 2\pi$. We may split the joint probability distribution into a distribution for R,

$$p_R(R) = Re^{-R^2/2}, \tag{C.5}$$

and a distribution for ϕ,

$$p_\phi(\phi) = \frac{1}{2\pi}. \tag{C.6}$$

The corresponding cumulative probability for R is

$$P_R(R) = \int_{-\infty}^{R} p_R(R) = 1 - e^{-R^2/2}, \tag{C.7}$$

which may be inverted to give

$$R = \sqrt{-2\ln(1 - P_R)}. \tag{C.8}$$

To generate a random R, we then simply generate a random number r_1 and set $R = \sqrt{-2\ln(r)}$. To generate a random ϕ, we generate a second random number r_2

1) Why not $x = -\ln(1 - r)$? This is because if r is uniformly distributed on the unit interval, so will $1 - r$ be. Hence, to save a subtraction for each random number, we simply rename $1 - r$ by r and generate these directly.

and set $\phi = 2\pi r_2$. We combine these results and transform back to the original x and y coordinates,

$$x = \sqrt{-2\ln(r_1)}\cos(2\pi r_2) \tag{C.9}$$

and

$$y = \sqrt{-2\ln(r_1)}\sin(2\pi r_2). \tag{C.10}$$

The numbers x and y will be distributed according to the Gaussian (C.3).

C.1.3
None of the Above

Finally, what if the distribution is none of the above, neither Gaussian nor having a cumulative probability $P(x)$ that can be inverted? In this case, one may resort to the von Neumann *rejection method,* [106].

Start by finding an auxiliary probability distribution $f(x)$ such that $f(x) > p(x)$ for all x but which has a corresponding cumulative probability $F(x) = \int_{-\infty}^{x} \mathrm{d}x' f(x')$, which is invertible. Let $f(x)$ be as close as possible to $p(x)$.

Generate a number x distributed according to $f(x)$. That is, from a random number r_1, find $x = F^{-1}(r_1)$. Generate a second number y from a distribution, which is uniform on the interval $0 \le y \le f(x)$. That is, generate a second random number r_2 and from this calculate $y = f(x)r_2$. Then,

- if $y \le p(x)$ accept x, or
- if $y > p(x)$, reject x.

The effectiveness of this algorithm depends on how close one is able to approximate $p(x)$ by $f(x)$.

We have one last advice to the reader concerning this topic: be careful with the random number generator you are using. There may be subtle correlations that can bias the results. It is always a good idea to run simulations with two different random number generators and compare the results. If the deviations are larger than the statistical fluctuations would imply, warning lights should start flashing.

C.2
Fourier Acceleration

The soft clamp fiber bundle model, described in Section 4.5 is computationally demanding. Here, we describe a technique that significantly accelerates the calculation of the force distribution among the fibers in this model. Even though we describe the technique in this specific context here, it is not limited to it. As the general ideas behind it are simple, it is not difficult to transcribe it to other systems.

The soft clamp fiber bundle model is defined by two equation sets (4.104)

$$\sigma_i = -k(u_i - x),$$

and (4.105),

$$u_i = \sum_j G_{ij}\sigma_j,$$

where the elastic Green function, G_{ij} is given in (4.106)

$$G_{ij} = \frac{1-\nu^2}{\pi E a^2}\int_{-a/2}^{a/2} \frac{dx'\,dy'}{|\vec{r}_i(x,y) - \vec{r}_j(x',y')|}.$$

We implement the model on a $L \times L = N$ square lattice.

By combining equations (4.104) and (4.105), we find the following equation:

$$(\mathbf{I} + \mathbf{KG})\vec{\sigma} = \mathbf{K}\vec{x}, \tag{C.11}$$

where we have introduced a matrix-vector notation. **I** is the $N \times N$ identity matrix, and **G** is the Green function represented as an $N \times N$ dense matrix. The constant vector $\vec{x}$ is N dimensional. The *diagonal* matrix **K** is also $N \times N$. Its matrix elements are either 1 for unbroken fibers or 0 for broken ones. The matrices **K** and **G** do not commute.

Once equation (C.11) is solved for the force the fibers carry, $\vec{\sigma}$, equation (4.105) easily yields the deformations of the elastic clamp.

Equation (C.11) is of the familiar form $\mathbf{A}\vec{\sigma} = \vec{b}$, where $\mathbf{A} = \mathbf{I} + \mathbf{KG}$ and $\vec{b} = \mathbf{K}\vec{x}$.

The simulation now proceeds as follows: we start with all intact fibers, each with its assigned failure threshold. The two media are then pulled apart, the forces calculated using the conjugate gradient algorithm [106, 107], and the fiber that is nearest to its threshold is broken, *i.e.* the matrix element corresponding to it in the matrix **K** is zeroed. Then, the new forces are calculated, a new fiber is broken, and so on until all fibers have failed.

However, there are two problems that render the simulation of large systems extremely difficult. The first is that since **G** is $N \times N$ *dense* matrix as it connects all fibers to all other fibers, the number of operations per conjugate gradient iteration scales like N^2. Even more serious is the fact that as the system evolves and springs are broken, the matrix $(\mathbf{I} + k\mathbf{G})$ becomes very ill-conditioned.

To overcome the problematic N^2 scaling of the algorithm, we note that the Green function is diagonal in Fourier space. Consequently, performing matrix-vector multiplications using fast Fourier transforms (FFT), the scaling is much improved and goes as $N^2 \ln(N)$. Symbolically, this can be expressed as follows:

$$(\mathbf{I} + \mathbf{KF}^{-1}\mathbf{FG})\mathbf{F}^{-1}\mathbf{F}\vec{\sigma} = \mathbf{K}\vec{x}, \tag{C.12}$$

where **F** is the FFT operator and $\mathbf{F}^{-1}$ its inverse ($\mathbf{F}^{-1}\mathbf{F} = 1$). Since **I** and **K** are diagonal, operations involving them are performed in real space. With this formulation, the number of operations/iterations in the conjugate gradient algorithm now scales as $N \ln(N)$.

To overcome the runaway behavior due to ill-conditioning, we need to precondition the matrix [107, 108]. This means that instead of solving Eq. (C.12), we solve the equivalent problem

$$\mathbf{Q}(\mathbf{I} + \mathbf{KF}^{-1}\mathbf{FG})\mathbf{F}^{-1}\mathbf{F}\vec{\sigma} = \mathbf{QK}\vec{x}, \tag{C.13}$$

where we have simply multiplied both sides by the arbitrary, positive definite preconditioning matrix $\mathbf{Q}$. Clearly, the ideal choice is $\mathbf{Q}_0 = (\mathbf{I} + \mathbf{KG})^{-1}$, which would always solve the problem in one iteration. Since this is not possible in general, we look for a form for $\mathbf{Q}$ that satisfies the following two conditions: (1) as close as possible to $\mathbf{Q}_0$ and (2) fast to calculate. The choice of a good $\mathbf{Q}$ is further complicated by the fact that as the system evolves and fibers are broken, corresponding matrix elements of $\mathbf{K}$ are set to zero. So, the matrix $(\mathbf{I} + \mathbf{KG})$ evolves from the initial form $(\mathbf{I} + \mathbf{G})$ to the final one $\mathbf{I}$.

A good choice for $\mathbf{Q}$ is

$$\mathbf{Q} = \mathbf{I} - (\mathbf{KG}) + (\mathbf{KG})(\mathbf{KG}) - (\mathbf{KG})(\mathbf{KG})(\mathbf{KG}) + \ldots \tag{C.14}$$

which is the Taylor series expansion of $\mathbf{Q}_0 = (\mathbf{I} + \mathbf{KG})^{-1}$. For best performance, the number of terms in the expansion is retained as a parameter since it depends on the physical parameters of the system. It is important to emphasize the following points. (a) As fibers are broken, the preconditioning matrix evolves with the ill-conditioned matrix and, therefore, remains a good approximation of its inverse throughout the breaking process. (b) All matrix multiplications involving $\mathbf{G}$ are done using FFTs. (c) The calculation of $\mathbf{Q}$ can be easily organized so that it scales like $nL^2 \ln(L)$, where n is the number of terms in the Taylor expansion, Eq. (C.14). The result is a stable Fourier-accelerated algorithm, which scales essentially as the volume of the system.

References

1. Lawn, B.R. (1993) *Fracture of Brittle Solids*, Cambridge University Press.
2. Buehler, M.J. (2008) *Atomistic Modeling of Materials Failure*, Springer-Verlag.
3. Hansen, A. (2014) *Front. Phys.*, **2**, 58.
4. Scholz, C.H. (2002) *The Mechanics of Earthquake and Faulting*, Cambridge University Press, Cambridge.
5. Sessions, R. (1950) *How a 'Difficult' Composer Gets that Way*, ProQuest, New York.
6. Peirce, F.T. (1926) *J. Text. Ind.*, **17**, 355.
7. Sornette, D. (1989) *J. Phys. A*, **22**, L243.
8. Lund, J.R. and Byrne, J.P. (2001) *Civ. Eng. Environ. Syst.*, **18**, 243.
9. Daniels, H.E. (1945) *Proc. R. Soc. London, Ser. A*, **183**, 405.
10. Pradhan, S., Hansen, A., and Chakrabarti, B.K. (2010) *Rev. Mod. Phys.*, **82**, 499.
11. Mallick, P.K. (1993) *Fiber Reinforced Composites: Materials, Manufacturing, and Design*, Marcel Dekker.
12. Cox, H.L. (1952) *Br. J. Appl. Phys.*, **3**, 72.
13. Hedgepeth, J.M. (1961) Stress Concentrations in Filamentary Structures. *NASA Technical Report* TND-882, NASA.
14. Mishnaevsky, L. (2007) *Computational Mesomechanics of Composites*, John Wiley & Sons, Ltd.
15. Domański, Z., Derda, T., and Sczygiol, N. (2013) *Proceedings of the International MultiConference of Engineers and Computer Scientists*, IMECS 2013, Vol. II.
16. Fjær, E. and Nes, O.M. (2014) Rock Mechanics and Rock Engineering, doi: 10.1007/s00 603–014–0598–5.
17. Weibull, W. (1951) *ASME J. Appl. Mech.*, **18**, 293.
18. Kloster, M., Hansen, A., and Hemmer, P.C. (1997) *Phys. Rev. E*, **56**, 2615.
19. Phoenix, S.L. and Taylor, H.M. (1973) *Adv. Appl. Probab.*, **5**, 200.
20. Barbour, A.D. (1975) *J. R. Stat. Soc. B*, **37**, 459.
21. Daniels, H.E. and Skyrme, T.H.R. (1985) *Adv. Appl. Probab.*, **17**, 85.
22. Hemmer, P.C. and Hansen, A. (1992) *ASME J. Appl. Mech.*, **59**, 909.
23. Aker, E., Måløy, K.J., Hansen, A., and Basak, S. (2000) *Europhys. Lett.*, **51**, 55–61.
24. Sornette, D. (1992) *J. Phys. I*, **2**, 2089.
25. Hansen, A. and Hemmer, P.C. (1994) *Trends Stat. Phys.*, **1**, 213.
26. Hemmer, P.C., Hansen, A., and Pradhan, S. (2007) *Modelling Critical and Catastrophical Phenomena in Geoscience: A Statistical Physics Approach*, Lecture Notes in Physics, Vol. 705, Springer-Verlag.
27. Feller, W. (1966) *Introduction to Probability Theory and Its Applications*, John Wiley & Sons, Inc.
28. Pradhan, S. and Hemmer, P.C. (2008) *Phys. Rev. E*, **77**, 031 138.
29. Pradhan, S., Bhattacharyya, P., and Chakrabarti, B.K. (2002) *Phys. Rev. E*, **66**, 016 116.
30. Hemmer, P.C. and Pradhan, S. (2007) *Phys. Rev. E*, **75**, 046 101.
31. Harlow, D.G. and Phoenix, S.L. (1991) *J. Mech. Phys. Solids*, **39**, 173.

The Fiber Bundle Model: Modeling Failure in Materials,
First Edition. Alex Hansen, Per C. Hemmer and Srutarshi Pradhan.

32. Gumbel, E.J. (1954) Statistical Theory of Extreme Values and Some Practical Applications, *National Bureau of Standards Applied Mathematics Series* **33**; (1958) *Statistics of Extremes*, Columbia University Press, New York (reprinted in 2013 by Echo Point Books & Media, Brattleborow).
33. Zhang, S.D. and Ding, E.J. (1994) *Phys. Lett. A*, **193**, 425.
34. Hansen, A. and Hemmer, P.C. (1994) *Phys. Lett. A*, **184**, 394.
35. Mahesh, S., Phoenix, S.L., and Beyerlein, I.J. (2002) *Int. J. Fract.*, **115**, 41.
36. Sinha, S., Kjellstadli, J.T., and Hansen, A. (2015) arXiv:1501.02489.
37. Wilkinson, D. and Willemsen, J.F. (1983) *J. Phys. A*, **16**, 3365.
38. Sahimi, M. and Arbabi, S. (1992) *Phys. Rev. Lett.*, **68**, 608.
39. Zapperi, S., Nukula, P., and Simunovic, S. (2005) *Phys. Rev. E*, **71**, 026 106.
40. Patinet, S., Vandembroucq, D., Hansen, A., and Roux, S. (2014) *Eur. J. Phys. Spec. Top.*, **223**, 2339.
41. Delaplace, A., Roux, S., and Pijaudier-Cabot, G. (2001) *J. Eng. Mech.*, **127**, 646.
42. Hidalgo, R.C., Kun, F., and Herrmann, H.J. (2002) *Phys. Rev. E*, **65**, 046 148.
43. Pradhan, S., Chakrabarti, B.K., and Hansen, A. (2005) *Phys. Rev. E*, **71**, 036 149.
44. Batrouni, G.G., Hansen, A., and Schmittbuhl, J. (2002) *Phys. Rev. E*, **65**, 036 126.
45. Stormo, A., Gjerden, K.S., and Hansen, A. (2012) *Phys. Rev. E*, **86**, R025 101.
46. Dugdale, D.S. (1960) *J. Mech. Phys. Solids*, **8**, 100.
47. Barenblatt, G.I. (1962) *Adv. Appl. Mech.*, **7**, 55.
48. Love, A.E.H. (1929) *Philos. Trans. R. Soc. London, Ser. A*, **228**, 377.
49. Johnson, K.L. (1985) *Contact Mechanics*, Cambridge University Press.
50. Landau, L.D. and Lifshits, E.M. (1986) *Theory of Elasticity*, Elsevier.
51. Stormo, A. (2013) *Brittle to Quasi-Brittle Transition in the Soft Clamp Fiber Bundle Model*. PhD thesis, NTNU, Trondheim.
52. Schmittbuhl, J. and Måløy, K.J. (1997) *Phys. Rev. Lett.*, **78**, 3888.
53. Schmittbuhl, J., Roux, S., Vilotte, J.P., and Måløy, K.J. (1995) *Phys. Rev. Lett.*, **74**, 1787.
54. Delaplace, A., Schmittbuhl, J., and Måløy, K.J. (1999) *Phys. Rev. E*, **60**, 1337.
55. Rosso, A. and Krauth, W. (2002) *Phys. Rev. E*, **65**, R025 101.
56. Måløy, K.J., Santucci, S., Schmittbuhl, J., and Toussaint, R. (2006) *Phys. Rev. Lett.*, **96**, 045 501.
57. Bonamy, D., Santucci, S., and Ponson, L. (2008) *Phys. Rev. Lett.*, **101**, 045 501.
58. Santucci, S., Grob, M., Toussaint, R., Schmittbuhl, J., Hansen, A., and Måløy, K.J. (2010) *Europhys. Lett.*, **92**, 44 001.
59. Laurson, L., Santucci, S., and Zapperi, S. (2010) *Phys. Rev. E*, **81**, 046 116.
60. Tallakstad, K.T., Toussaint, R., Santucci, S., Schmittbuhl, J., and Måløy, K.J. (2011) *Phys. Rev. E*, **83**, 046 108.
61. Lengliné, O., Toussaint, R., Schmittbuhl, J., Elkhoury, J.E., Ampuero, J.P., Tallakstad, K.T., Santucci, S., and Måløy, K.J. (2011) *Phys. Rev. E*, **84**, 036 104.
62. Schmittbuhl, J., Hansen, A., and Batrouni, G.G. (2003) *Phys. Rev. Lett.*, **90**, 045 505.
63. Gjerden, K.S., Stormo, A., and Hansen, A. (2013) *Phys. Rev. Lett.*, **111**, 135 502.
64. Gjerden, K.S., Stormo, A., and Hansen, A. (2014) *Front. Phys.*, **2**, 66.
65. da Silveira, R. (1999) *Am. J. Phys.*, **67**, 1177.
66. Pradhan, S. and Chakrabarti, B.K. (2001) *Phys. Rev. E*, **65**, 016 113.
67. Pradhan, S. and Hemmer, P.C. (2007) *Phys. Rev. E*, **75**, 056 112.
68. Sahimi, M. (1994) *Applications of Percolation Theory*, Taylor & Francis Ltd, London.
69. Pradhan, S., Hansen, A., and Hemmer, P.C. (2005) *Phys. Rev. Lett.*, **95**, 125 501.

70. Gutenberg, B. and Richter, C.F. (1954) *Seismicity of the Earth and Associated Phenomena*, 2nd edn, Princeton University Press.
71. Kawamura, H., Hatano, T., Kato, N., Biswas, S., and Chakrabarti, B.K. (2012) *Rev. Mod. Phys.*, **84**, 839.
72. Smith, W.D. (1981) *Nature*, **289**, 136.
73. Latchman, J.L., Morgan, F.D.O., and Aspinal, W.P. (2008) *Earth Sci. Rev.*, **87**, 94.
74. Kawamura, H. (2006) arXiv:cond-mat/0603335.
75. Tsakakoshi, Y. and Shimazaki, K. (2008) *Earth, Planets Space*, **60**, 915.
76. Nanjo, K.Z., Hiraka, N., Obara, K., and Kasahara, K. (2012) *Geophys. Res. Lett.*, **39**, L020 304.
77. Chan, C.H., Wu, Y.M., Tzeng, T.L., Lin, T.L., and Chen, C.C. (2012) *Tectonophysics*, **532**, 215.
78. de Arcangelis, L., Redner, S., and Herrmann, H.J. (1985) *Phys. Rev. E*, **46**, L585.
79. Alava, M.J., Nukala, P.K.N.N., and Zapperi, S. (2006) *Adv. Phys.*, **55**, 349.
80. Pradhan, S., Hansen, A., and Hemmer, P.C. (2006) *Phys. Rev. E*, **74**, 016 122.
81. Petri, A., Paparo, G., Vespignani, A., Alippi, A., and Costantini, M. (1994) *Phys. Rev. Lett.*, **73**, 3423.
82. Pradhan, S. and Hemmer, P.C. (2011) *Phys. Rev. E*, **83**, 041 116.
83. Pradhan, S., Stroisz, A., Fjær, E., Stenebraten, J., Lund, H.K., Sønstebø, F.F., and Roy, S. (2014) ARMA-2014-7442.
84. Kun, F., Zapperi, S., and Herrmann, H.J. (2000) *Eur. Phys. J. B*, **17**, 269.
85. Suresh, S. (1991) *Fatigue of Materials*, Cambridge University Press.
86. Hidalgo, R.C., Kun, F., and Herrmann, H.J. (2001) *Phys. Rev. E*, **64**, 066 122.
87. Evans, G. and Zok, F.W. (1994) *J. Mater. Sci.*, **29**, 3857.
88. Kun, F., Costa, M.H., Costa Filho, R.N., Andrade, J.S. Jr., Soares, J.B., Zapperi, S., and Herrmann, H.J. (2007) *J. Stat. Mech.*, **P02003**, 10.
89. Kun, F., Carmona, H.A., Andrade, J.S. Jr., and Herrmann, H.J. (2008) *Phys. Rev. Lett.*, **100**, 094 301.
90. Leomach, M., Perge, C., Divoux, T., and Manneville, S. (2014) arXiv: 1401.8234v1.
91. Roux, S. (2000) *Phys. Rev. E*, **62**, 6164.
92. Guarino, A., Garcimartin, A., and Ciliberto, S. (1999) *Europhys. Lett.*, **47**, 456.
93. Pradhan, S. and Chakrabarti, B.K. (2003) *Phys. Rev. E*, **67**, 046 124.
94. Greer, J.R., Jang, D., Kim, J.Y., and Burek, M.J. (2009) *Adv. Funct. Mater.*, **19**, 2880.
95. Sievila, P., Chekurov, N., and Tittonen, I. (2010) *Nanotechnology*, **21**, 145 301.
96. Fjær, E. (1999) *Rock Mechanics for Industry* (eds Amadei, B., Kranz, R.L., Scott, G.A., and Smeallie, P.H.), A.A. Balkema.
97. Reiweger, I. (2011) *Failure of Weak Snow Layers*. PhD thesis, ETH, Zürich.
98. Reiweger, I., Schwarz, J., Dual, J., and Herrmann, H.J. (2009) *J. Glaciol.*, **55** (194), 997.
99. Jamieson, J.B. and Schweizer, J. (2000) *J. Glaciol.*, **46**, 151.
100. Lehmann, P. and Or, D. (2012) *Water Resour. Res.*, **48**, W03 535.
101. Cohen, D., Schwarz, M., and Or, D. (2011) *J. Geophys. Res.* **16**, F03010.
102. Bak, P., Tang, C., and Wiesenfeld, K. (1988) *Phys. Rev. A*, **38**, 364.
103. Whittaker, E.T. and Watson, G.N. (1958) *A Course of Modern Analysis*, Cambridge University Press.
104. Tennekes, H. and Lumley, J.L. (1972) *A First Course in Turbulence*, MIT Press.
105. David, H.A. (1981) *Order Statistics*, John Wiley & Sons, Inc.
106. Press, W.H., Teukolsky, S.A., Vetterling, W.T., and Flannery, B.P. (2007) *Numerical Recipes*, 3rd edn, Cambridge University Press.
107. Batrouni, G.G. and Hansen, A. (1988) *J. Stat. Phys.*, **52**, 747.
108. Batrouni, G.G., Hansen, A., and Nelkin, M. (1986) *Phys. Rev. Lett.*, **57**, 1336.

Index

a
acoustic emission 53
asphalt experiment 186
asymptotes 220–221
asymptotic burst distribution
– generic case 40–42
– nongeneric cases 42–45
– via random walks 51–53
– by stepwise load increase 57–59
avalanches
– at criticality 141–142
– as random walks 49–53
– size 33
– size distribution 138–142
– snow 197–199
– by stepwise load increases 57–61
average force 13, 28

b
backward condition 36–37, 45
Basquin exponent 186
Basquin's law 186
binomial coefficient 205
Box-Müller algorithm 224
breakdown susceptibility 119
breaking rule 64
Brownian motion 30, 211–212
– scaling 30
bundle strength 11
– local load sharing model, upper bound 82
– soft clamp model 110
burst avalanches 32
– equal-load-sharing model 14
burst distribution 138
– asymptotics 41, 53
– at criticality 141–142
– equal load sharing model 33–40
– local load sharing model 90–93, 99
bursts
– average size 38–40
– backward condition 34, 36–37
– forward condition 34–36
– size 27
– by stepwise load increase 57
– total number 37–38

c
central limit theorem 211, 215–216
change of variable 213–214
chaos theory 115
characteristic function 214–215
clamp
– rigid 12, 63, 137
– soft 12, 63, 107
– stiff 63
combinatorial theorems 204–206
– restricted distribution 205–206
– selection theorems 204–205
compressive loading 195
computational physics 6
conjugate gradient algorithm 226
coordinate transformation 223
corrosion 2
Cox shear lag model 8
creep 169
– damage accumulation 180–181
– deformation rate 175
– exponent 178
– failure 175–183
– healing 182–183
– model 176–183
– time evolution 181–182
critical dynamics 119–135
– amplitude ratio 135
– critical relaxation 124

The Fiber Bundle Model: Modeling Failure in Materials, First Edition.
Alex Hansen, Per C. Hemmer and Srutarshi Pradhan.

critical dynamics (*contd.*)
– critical slowing down 120
– general threshold distribution 124–125
– susceptibility 119
critical force 17
critical indices 115
critical phenomena 6, 115
critical point 18
– breakdown 18
critical stress 18
crossover phenomenon 138–152
– uniform distribution 140
cumulants, statistical 215
cumulative distribution function 11, 213

d
da Vinci, Leonardo 4
damage accumulation 180–181
Daniels, Henry 11, 15–16
dense matrix 226
density function 11
dynamical systems theory 115

e
earth quakes 144–146
elastic constant 11, 13, 22, 28
elastic medium anchoring 107–114
elastic–plastic model 25–26
elasticity
– linear 12
electromechanical systems 194–195
energy emission bursts 159–166
energy release 53–57, 142
– at high energies 55
– at low energies 55–57
equal load sharing model 11–12
– burst distribution 37
– forward bursts 47–48
– asymptotic burst distribution 40–42
– failure prediction 137
– hole size distribution 79–80
– inclusive bursts 45–46
– as mean field model 11
– recursive dynamics 116
explosive decompression 2
extreme statistics 218–221
extreme value distributions 218–221

f
failure mechanisms 85–86
failure time
– low noise 189–190
fast Fourier transforms 226
fatigue 2, 169, 186–188
– Basquin's law 186
– cyclic loading 186
– experiment 186–188
– failure 186
fiber mixtures 22–24
fiber reinforced composites 8, 170, 171
fiber strengths 11, 13
– strictly uniform distribution 16
– uniform distribution 13, 18
– Weibull distribution 18
fibers with internal structure 171
first asymptote 219
fluctuations 4, 27–61
force
– average 17
– macroscopic 17
force fluctuations 28–29
– range 28–29
force-controlled experiment 12–15, 152–154
forward bursts 47–48
Fourier acceleration 225–227
fracture front propagation 112–114
fuse model 6, 148–152
– burst distribution 150
– crossover 151
– current-voltage relation 149

g
γ model 105–106
Galilei, Galileo 4
Gaussian distribution 215
Gaussian numbers 224–225
glacier flow 175
global load sharing model 11
Green function 108
Gumbel, Emil Julius 73–74
Gumbel distribution 219
Gutenberg, Beno 147–148
Gutenberg–Richter law 144–145, 147

h
Hedgepeth model 8
history dependency 66
hole size distribution 78–79
– equal load sharing model 79–80
– local load sharing model 80–82

i
inclusive bursts
– equal load sharing model 45–46
– local load sharing model 92
– infinite disorder limit 75–76

intergranular failure 195
intermediate load sharing models 104–106
intragranular failure 195
invasion percolation 95
iterated maps 115

l
Lagrange's inversion theorem 203–204
landslide 199
– model 200
– shallow landslides 199–201
– triggering 200
least monotonic function (LMF) 14
Liberty ships 2
linear elastic fracture mechanics (LEFM) 1
load 11
– redistribution 11
load curve 13
– generic case 30
– generic form 21
– with no maximum 21
– with non-parabolic maximum 21
– with several parabolic maxima 21
– uniform distribution 20
– Weibull distribution 20
– work hardening 171
local load sharing model 12, 63
– bundle strength 68–69
– competing failure mechanisms 85–86
– failure of first and second fiber 69–78
– failure sequence 67–68
– force and elongation characteristics 88–90
– hole size distribution 78–82
– localization 76–78
– in one dimension 65
– redistribution of forces 66–67
– two dimensions 93–101
three and four dimensions 100–101
localization 76–78, 80–82, 94–95
– soft clamp model 109–110

m
macroscopic force 17
Mariotte, Edme 4
maximum bundle strength 30–32
maximum force 17
mean field theory 115
mechanical energy in fiber 53
mixed-mode model 106
models 3–4
– hierarchical approach 3–4
moments, statistical 214

n
nanopillars 194–195
noise
– lifetime 191
– waiting time 192
noise-induced failure 190–194
non-hookean forces 24–26
nonlinear dynamics 115, 117

o
Occam's razor 4
order statistics 71, 216–217
– averages 217
– joint probabilities 217–221

p
Peirce, Fredrick Thomas 5
pillar model 8, 195
Poisson process 34
Poisson ratio 109
porous rocks 195
postcritical relaxation
– uniform distribution 126–128
– Weibull distribution 129–130
precritical relaxation
– uniform distribution 131–133
– Weibull distribution 133–134
predicting failure 137–168
– via breaking rate 155
– burst size divergence 152
– via energy emission 159–166

q
quasistatic loading 67

r
random numbers
– Gaussian numbers 224–225
– generation 223–225
random slacks 24–25
random walk 206–208
– asymmetrical 209–211
– biased 206–208
– Brownian motion 211–212
– gambler's ruin 207–208
– probability of no return 206–207
– Wiener process 211
recursive dynamics 115
– critical dynamics 119–135
– fixed points 116–119
– linearly increasing distribution 120
– order parameter 119, 123
– postcritical relaxation 125–130
– precritical relaxation 131–134

recursive dynamics (*contd.*)
– uniform distribution 118
– universal amplitude ratio 134–135
– universal critical behavior 120–125
– Weibull distribution 129
relaxation time 154
repeated damage 171–174
– complete failure 174–175
Richter, Charles 146–148
Richter's scale 147
roots 199, 201

s
sandpile model 200
second asymptote 221
shallow landslides 199–201
slab snow avalanches 197–199
snow avalanches 197–199
– sintering 198
soft clamp model 107, 225
soft membrane model 101–104
spatial correlations 11, 115
stochastic variable 213–216
strain hardening 169
strain-controlled experiment 13
stress 13
support 11
– elastic medium 12
– non-rigid 12
– rigid 11
– soft 12
susceptibility 154

t
theory of elasticity 1
thermally induced failure 188–190
three extreme value asymptotes 218–221
thresholds 13
– strictly uniform distribution 16
– uniform distribution 13
– Weibull distribution 18

u
uniform threshold distribution 18
upper critical dimension 99–101

v
viscoelastic creep 169, 183–186

w
weak layer 197
weakest link of a chain 221
Weibull, Waloddi 19–20
Weibull distribution 19, 221
Weibull index 18
work hardening 169–175
– load curve 171
– plateau value 174
– repeated damage 171–174

y
Young modulus 109